NEUROMETHODS

C000104127

Series Editor
Wolfgang Walz
University of Saskatchewan
Saskatoon, SK, Canada

For further volumes:
http://www.springer.com/series/7657

Neuromethods publishes cutting-edge methods and protocols in all areas of neuroscience as well as translational neurological and mental research. Each volume in the series offers tested laboratory protocols, step by step methods for reproducible lab experiments and addresses methodological controversies and pitfalls in order to aid neuroscientists in experimentation. *Neuromethods* focuses on traditional and emerging topics with wide ranging implications to brain function, such as electrophysiology, neuroimaging, behavioral analysis, genomics, neurodegeneration, translational research and clinical trials. *Neuromethods* provides investigators and trainees with highly useful compendiums of key strategies and approaches for successful research in animal and human brain function including translational "bench to bedside" approaches to mental and neurological diseases.

Neuroproteomics

Second Edition

Edited by

Ka Wan Li

Department of Molecular and Cellular Neurobiology, Center for Neurogenomics and Cognitive Research, Amsterdam Neuroscience, Vrije Universiteit Amsterdam, Amsterdam, The Netherlands

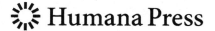 Humana Press

Editor
Ka Wan Li
Department of Molecular and Cellular Neurobiology
Center for Neurogenomics and Cognitive
Research, Amsterdam Neuroscience
Vrije Universiteit Amsterdam
Amsterdam, The Netherlands

ISSN 0893-2336   ISSN 1940-6045   (electronic)
Neuromethods
ISBN 978-1-4939-9664-3   ISBN 978-1-4939-9662-9   (eBook)
https://doi.org/10.1007/978-1-4939-9662-9

This Humana imprint is published by the registered company Springer Science+Business Media, LLC, part of Springer Nature.
The registered company address is: 233 Spring Street, New York, NY 10013, U.S.A.

Preface to the Series

Experimental life sciences have two basic foundations: concepts and tools. The *Neuromethods* series focuses on the tools and techniques unique to the investigation of the nervous system and excitable cells. It will not, however, shortchange the concept side of things as care has been taken to integrate these tools within the context of the concepts and questions under investigation. In this way, the series is unique in that it not only collects protocols but also includes theoretical background information and critiques which led to the methods and their development. Thus it gives the reader a better understanding of the origin of the techniques and their potential future development. The *Neuromethods* publishing program strikes a balance between recent and exciting developments like those concerning new animal models of disease, imaging, in vivo methods, and more established techniques, including, for example, immunocytochemistry and electrophysiological technologies. New trainees in neurosciences still need a sound footing in these older methods in order to apply a critical approach to their results.

Under the guidance of its founders, Alan Boulton and Glen Baker, the *Neuromethods* series has been a success since its first volume published through Humana Press in 1985. The series continues to flourish through many changes over the years. It is now published under the umbrella of Springer Protocols. While methods involving brain research have changed a lot since the series started, the publishing environment and technology have changed even more radically. *Neuromethods* has the distinct layout and style of the Springer Protocols program, designed specifically for readability and ease of reference in a laboratory setting.

The careful application of methods is potentially the most important step in the process of scientific inquiry. In the past, new methodologies led the way in developing new disciplines in the biological and medical sciences. For example, Physiology emerged out of Anatomy in the nineteenth century by harnessing new methods based on the newly discovered phenomenon of electricity. Nowadays, the relationships between disciplines and methods are more complex. Methods are now widely shared between disciplines and research areas. New developments in electronic publishing make it possible for scientists that encounter new methods to quickly find sources of information electronically. The design of individual volumes and chapters in this series takes this new access technology into account. Springer Protocols makes it possible to download single protocols separately. In addition, Springer makes its print-on-demand technology available globally. A print copy can therefore be acquired quickly and for a competitive price anywhere in the world.

Saskatoon, SK, Canada   *Wolfgang Walz*

Preface

The field of proteomics evolves rapidly. The first edition of *Neuroproteomics* from a decade ago described various aspects of experimental designs for proper proteomics experiments. Today, we are approaching the analysis of tissue/cellular proteins at near proteome-wide level. New sample preparation protocols have been developed. Mass spectrometers are achieving higher sensitivity and speed, resolution, and mass accuracy, which enable new MS data acquisition methods with increased reproducibility and protein coverage. In view of the substantial improvements in various aspects of proteomics, we launch a second edition to provide readers with protocols most used in today's proteomics workflows.

The present edition has 16 chapters, 14 of which cover aspects of various dimensions of the new proteomics technology, and two chapters are updated versions from the previous edition. These protocols can serve as a guide to obtain the best results from current proteomics platforms.

Amsterdam, The Netherlands *Ka Wan Li*

Contents

Contributors

KRISHNA D. B. ANAPINDI • *Department of Chemistry, Beckman Institute, University of Illinois, Urbana, IL, USA*

DAVID AVILA • *Roche Pharma Research and Early Development, Pharmaceutical Sciences, Roche Innovation Center Basel, F. Hoffmann-La Roche Ltd., Basel, Switzerland*

KAJ BLENNOW • *Department of Psychiatry and Neurochemistry, Mölndal Hospital, Sahlgrenska Academy at the University of Gothenburg, Mölndal, Sweden; Clinical Neurochemistry Laboratory, Sahlgrenska University Hospital Mölndal, Mölndal, Sweden*

NING CHEN • *State Key Laboratory of Proteomics, Beijing Proteome Research Center, National Center for Protein Sciences Beijing, Beijing Institute of Lifeomics, Beijing, China*

TOM DUNKLEY • *Roche Pharma Research and Early Development, Pharmaceutical Sciences, Roche Innovation Center Basel, F. Hoffmann-La Roche Ltd., Basel, Switzerland*

LEONIDAS FALIAGKAS • *Department of Molecular and Cellular Neurobiology, Center for Neurogenomics and Cognitive Research, Amsterdam Neuroscience, Vrije Universiteit Amsterdam, Amsterdam, The Netherlands*

TITIA GEBUIS • *Department of Molecular and Cellular Neurobiology, Faculty of Science, Center for Neurogenomics and Cognitive Research, Vrije Universiteit Amsterdam, Amsterdam, The Netherlands*

MIGUEL A. GONZALEZ-LOZANO • *Department of Molecular and Cellular Neurobiology, Center for Neurogenomics and Cognitive Research, Amsterdam Neuroscience, Vrije Universiteit Amsterdam, Amsterdam, The Netherlands*

JÖRG HANRIEDER • *Department of Psychiatry and Neurochemistry, Mölndal Hospital, Sahlgrenska Academy at the University of Gothenburg, Mölndal, Sweden; Department of Neurodegenerative Disease, Queen Square Institute of Neurology, University College London, London, UK*

FUCHU HE • *State Key Laboratory of Proteomics, Beijing Proteome Research Center, National Center for Protein Sciences Beijing, Beijing Institute of Lifeomics, Beijing, China*

DAVID C. HONDIUS • *Department of Molecular and Cellular Neurobiology, Center for Neurogenomics and Cognitive Research, Amsterdam Neuroscience, Vrije Universiteit Amsterdam, Amsterdam, The Netherlands; Department of Pathology, Amsterdam Neuroscience, Amsterdam UMC, Amsterdam, The Netherlands*

JEROEN J. M. HOOZEMANS • *Department of Pathology, Amsterdam Neuroscience, Amsterdam UMC, Amsterdam, The Netherlands*

RAVI JAGASIA • *Roche Pharma Research and Early Development, Neuroscience and Rare Diseases Discovery and Translational Medicine Area, Roche Innovation Center Basel, F. Hoffmann-La Roche Ltd., Basel, Switzerland*

CONNIE R. JIMENEZ • *OncoProteomics Laboratory, Department of Medical Oncology, Cancer Center Amsterdam, Amsterdam UMC, Amsterdam, The Netherlands*

EIJI KINOSHITA • *Department of Functional Molecular Science, Institute of Biomedical and Health Sciences, Hiroshima University, Hiroshima, Japan*

EMIKO KINOSHITA-KIKUTA • *Department of Functional Molecular Science, Institute of Biomedical and Health Sciences, Hiroshima University, Hiroshima, Japan*

TOHRU KOIKE • *Department of Functional Molecular Science, Institute of Biomedical and Health Sciences, Hiroshima University, Hiroshima, Japan*

FRANK KOOPMANS • *Department of Molecular and Cellular Neurobiology, Center for Neurogenomics and Cognitive Research, Amsterdam Neuroscience, Vrije Universiteit Amsterdam, Amsterdam, The Netherlands*

KA WAN LI • *Department of Molecular and Cellular Neurobiology, Center for Neurogenomics and Cognitive Research, Amsterdam Neuroscience, Vrije Universiteit Amsterdam, Amsterdam, The Netherlands*

WEIDONG LI • *Bio-X Institutes, Key Laboratory for the Genetics of Development and Neuropsychiatric Disorders (Ministry of Education), Shanghai Key Laboratory of Psychotic Disorders, Brain Science and Technology Research Center, Institute of psychology and Behavioral Sciences, Shanghai Jiao Tong University, Shanghai, China; Institute for Biomedical Sciences, Interdisciplinary Cluster for Cutting Edge Research, Shinshu University, Matsumoto, Japan*

FAN LIU • *Leibniz-Forschungsinstitut für Molekulare Pharmakologie (FMP), Berlin, Germany*

MINGWEI LIU • *State Key Laboratory of Proteomics, Beijing Proteome Research Center, National Center for Protein Sciences Beijing, Beijing Institute of Lifeomics, Beijing, China*

IRYNA PALIUKHOVICH • *Department of Molecular and Cellular Neurobiology, Faculty of Science, Center for Neurogenomics and Cognitive Research, Vrije Universiteit Amsterdam, Amsterdam, The Netherlands*

NIKHIL J. PANDYA • *Department of Molecular and Cellular Neurobiology, Center for Neurogenomics and Cognitive Research, Amsterdam Neuroscience, Vrije Universiteit Amsterdam, Amsterdam, The Netherlands; Roche Pharma Research and Early Development, Pharmaceutical Sciences, Roche Innovation Center Basel, F. Hoffmann-La Roche Ltd, Basel, Switzerland*

THANG V. PHAM • *OncoProteomics Laboratory, Department of Medical Oncology, Cancer Center Amsterdam, Amsterdam UMC, Amsterdam, The Netherlands*

JUN QIN • *State Key Laboratory of Proteomics, Beijing Proteome Research Center, National Center for Protein Sciences Beijing, Beijing Institute of Lifeomics, Beijing, China*

PRIYANKA RAO-RUIZ • *Department of Molecular and Cellular Neurobiology, Center for Neurogenomics and Cognitive Research, Amsterdam Neuroscience, Vrije Universiteit Amsterdam, Amsterdam, The Netherlands*

ELENA V. ROMANOVA • *Department of Chemistry, Beckman Institute, University of Illinois, Urbana, IL, USA*

MARTINA ROSATO • *Department of Molecular and Cellular Neurobiology, Faculty of Science, Center for Neurogenomics and Cognitive Research, Vrije Universiteit Amsterdam, Amsterdam, The Netherlands*

ANNEMIEKE J. M. ROZEMULLER • *Department of Pathology, Amsterdam Neuroscience, Amsterdam UMC, Amsterdam, The Netherlands*

YOSHINORI SHIRAI • *Department of Molecular and Cellular Physiology, Institute of Medicine, Shinshu University Academic Assembly, Matsumoto, Japan*

AUGUST B. SMIT • *Department of Molecular and Cellular Neurobiology, Faculty of Science, Center for Neurogenomics and Cognitive Research, Amsterdam Neuroscience, Vrije Universiteit Amsterdam, Amsterdam, The Netherlands*

SABINE SPIJKER • *Department of Molecular and Cellular Neurobiology, Center for Neurogenomics and Cognitive Research, Amsterdam Neuroscience, Vrije Universiteit Amsterdam, Amsterdam, The Netherlands*

SVEN STRINGER • *Department of Complex Trait Genetics, Faculty of Science, Center for Neurogenomics and Cognitive Research, Vrije Universiteit Amsterdam, Amsterdam, The Netherlands*

PATRICK F. SULLIVAN • *Department of Medical Epidemiology and Biostatistics, Karolinska Institutet, Stockholm, Sweden; Department of Genetics, University of North Carolina, Chapel Hill, NC, USA*

WEI SUN • *State Key Laboratory of Proteomics, Beijing Proteome Research Center, National Center for Protein Sciences Beijing, Beijing Institute of Lifeomics, Beijing, China*

TATSUO SUZUKI • *Department of Molecular and Cellular Physiology, Institute of Medicine, Shinshu University Academic Assembly, Matsumoto, Japan*

JONATHAN V. SWEEDLER • *Department of Chemistry, Beckman Institute, University of Illinois, Urbana, IL, USA*

MANUEL TZOUROS • *Roche Pharma Research and Early Development, Pharmaceutical Sciences, Roche Innovation Center Basel, F. Hoffmann-La Roche Ltd., Basel, Switzerland*

SOPHIE J. F. VAN DER SPEK • *Department of Molecular and Cellular Neurobiology, Center for Neurogenomics and Cognitive Research, Amsterdam Neuroscience, Vrije Universiteit Amsterdam, Amsterdam, The Netherlands*

RONALD E. VAN KESTEREN • *Department of Molecular and Cellular Neurobiology, Faculty of Science, Center for Neurogenomics and Cognitive Research, Vrije Universiteit Amsterdam, Amsterdam, The Netherlands*

FLORIAN WEILAND • *Medical Research Council Protein Phosphorylation and Ubiquitylation Unit, Sir James Black Centre, School of Life Sciences, University of Dundee, Dundee, UK*

FANG XIE • *Department of Chemistry, Beckman Institute, University of Illinois, Urbana, IL, USA*

HENRIK ZETTERBERG • *Department of Psychiatry and Neurochemistry, Mölndal Hospital, Sahlgrenska Academy at the University of Gothenburg, Mölndal, Sweden; Department of Neurodegenerative Disease, Queen Square Institute of Neurology, University College London, London, UK; UK Dementia Research Institute at UCL, London, UK; Clinical Neurochemistry Laboratory, Sahlgrenska University Hospital Mölndal, Mölndal, Sweden*

## Chapter 1

Neuroproteomics: The Methods

Ka Wan Li

Abstract

Neuroproteomics is a branch of proteomics that analyses the (sub-)proteomes of the nervous system qualitatively or quantitatively. This chapter introduces the various methodologies that are commonly used in neuroproteomics research.

Key words Nervous system, Proteomics, Mass spectrometer

1 Introduction

The brain is the most complex organ in our body with billions of neurons that communicate within the CNS and to the periphery mainly via hundreds of trillions of synapses engaged in neurotransmission. A diversity of neuronal cellular activities underlie how we perceive external sensory signals and internal cognitive/physiological states, process and integrate the information, and produce appropriate outputs accordingly. Ultimately, the specific spatiotemporal activities and their functional interaction within this huge neuronal network define our personality. Pathological disturbance to synaptic transmission and the alteration of cellular functions are at the basis of many brain disorders, for example, psychiatric disorders, such as depression, schizophrenia, and drug addiction, and neurodegenerative diseases including the highly prevalent Alzheimer's and Parkinson's diseases in the aged population [1, 2].

The understanding of cellular and synaptic functions at the molecular level has been an important aspect of neuroscience research. In the past decade proteomics studies have revealed the constituents of the synapse to amount up to 2000–5000 proteins [3–5]. Quantitative proteomics have been applied to detect the dynamic changes in the synaptic proteome, and in some cases these changes could be causally related to physiological/behavioural changes [6–8]. More recently, the complex human brain proteome started to be examined, in which spatial and temporal alteration of distinct functional proteins or protein groups from degenerative disease patients have been demonstrated [9–13]. Proteomics has become a tool that is expected to lead to new discoveries in the human brain in health and disease in the years to come.

2   Proteomics Methods, Step by Step

Neuroproteomics is a powerful analytical method that can help to understand functioning and the disorders of the nervous systems from the global protein expression perspective. A fundamental consideration is the selection of biologically relevant samples, in particular the brain regions of interest. In Chapter 2, a comprehensive protocol to dissect mouse brain regions is described. The isolated brain region can then be analysed directly as total lysate. Studies may further focus on subcellular domains, such as the synapse. The lower protein complexity of this and other organelles allows a deeper and more complete analysis of its proteome. Chapter 3 describes the chemical fractionation of synaptic subdomains down to the post-synaptic density and the recently discovered post-synaptic density lattice [14]. Using biochemical separation only, a particular brain region remains heterogeneous with regard to subregions that may execute different functions, and contain different protein constituents. For example, hippocampus can be distinguished into different CA regions and dentate gyrus. Within a subregion there are obviously also different cell types including neurons, astrocytes, microglia that may in turn react differently to stimuli or pathological insults. Therefore, it may be necessary to focus on specific brain subregions and even single cell types. Laser microdissection microscopy [15] is the method of choice for high spatial resolution dissection and collection of user-defined tissue area down to the single cell level. It has been applied successfully to the proteomics study of neurodegenerative disorders from postmortem human brain samples [13, 16, 17]. This methodology is described in Chapter 4.

The thousands of cellular proteins intrinsically have diverse biophysical properties, for example, observed in the hydrophobic membrane proteins, the hydrophilic cytosolic proteins, and possibly some cellular protein aggregates that form insoluble particles. A bottom-up proteomics workflow requires solubilization of cellular proteins and protease digestion before downstream liquid chromatography-tandem mass spectrometry (LC-MS/MS) analysis of the resulting peptides. Several protocols have been developed that are optimal for protein solubilization without compromising the subsequent protease digestion, and in which the peptides are recovered in a condition that is compatible to LC-MS/MS analysis. One popular workflow involves the use of SDS-PAGE. Sample is solubilized in SDS sample buffer, proteins are then partial size fractionated by PAGE and subsequently visualized by staining with a dye such as Coomassie Blue. The destaining-dehydration-rehydration step of the gel pieces serves as a cleaning step that removes small molecular weight contaminants including SDS. This procedure was described in detail in a chapter in the previous edition of Neuroproteomics [18], and is integrated as a subsection in Chapters 4 and 11 in the present edition. The gel-based protocol is time-consuming and labour-intense. An in-solution digestion protocol is simpler to handle and can be scaled up for high-throughput sample preparation. Filter-aided sample preparation (FASP) has gain popularity in recent years [19]. All washing and digestion steps are carried out in a single filter unit. The removal of the successive washes from the filter and the collection of digested peptides are done by centrifugation. Another recently introduced method, the single-pot solid-phase sample preparation (SP3, [20]), has the possibility for large-scale analysis and may provide a means for ultra-high-sensitivity analysis. These two protocols are described in Chapter 6. While FASP and SP3 recover in principle all tryptic peptides, there are methods that capture peptides with specific posttranslational modifications (PTMs). Peptides with PTM, such as phosphorylation, often exist at low level, and require enrichment for their detection. Chapter 7 describes the phos-tag tip technology to enrich low molecular weight phosphorylated biomolecules including phosphopeptides.

Proteins seldom act in isolation. Most proteins interact with other proteins and are part of complexes to carry out physiological or pathophysiological processes. Interaction proteomics has been employed extensively to capture and identify protein complexes (for example [21–23]). Immuno-affinity isolation of protein complexes is incorporated as a subsection in Chapter 8. A protein may be present in multiple protein complexes with different functions. Conventional immunoprecipitation-based interaction proteomics fails to distinguish protein subcomplexes. In Chapter 8, Blue Native gel electrophoresis is introduced as an extra dimension to distinguish protein subcomplexes, provided that they differ sufficiently in mass by which they can be resolved on the gel.

To capture protein complexes by an antibody-based method, tissue must be prior solubilized by detergent to release individual protein complexes into solution. The caveat is that the stability of protein–protein interaction, or part of the protein complex, is strongly affected by the type of detergent used [24]. At the opposite of the spectrum, the detergent is not strong enough to solubilize the targeted protein complexes, which then precludes subsequent analysis. The recently developed protocol using a MS gas phase cleavable crosslinker addresses these problems from another angle [25, 26]. Proteins that reside in vicinity in vivo presumably belong to the same protein complex. They can be crosslinked under native conditions, for example, within a synaptosome. After protease digestion, peptide pairs can be selectively identified by LC-MS/MS based on the presence of the signature ions derived from the crosslinker, which in turn infer the interaction of the respective proteins. The protocol for protein crosslink-mass spectrometry is described in Chapter 9.

There are several MS-based protocols for quantitative proteomics. Tryptic peptides are loaded and partially separated by LC and analyzed by the mass spectrometer. The peak height or area of the peptides from MS1 can be used for quantitation, whereas the fragment ions generated from MS/MS of the selected precursor peptide serve to identify the peptide by database search (data dependent acquisition DDA). This is described in subsections of Chapters 5, 9, and 14. A variant of DDA is the use of isobaric multiplex labelling strategy [27]. Here, up to 11 samples (TMT), which are separately tagged by the 11 different isobaric tags, can be pooled and analyzed simultaneously. The same peptide from different samples is then quantified based on the intensity of the respective signature ions derived from the corresponding fragmented isobaric tags (Chapter 10). Fragment ions from MS/MS can also be used for quantitation. For global analysis, all peptides within a wide precursor ion selection window are fragmented simultaneously and quantified. The identities of the peptides are revealed by mapping of the fragment ions to those present in a spectral library. This data-independent acquisition or SWATH [28]) is described in Chapter 11. Finally, in Chapter 12, and subsections of Chapter 14, the various computational strategies for the analysis of DDA and DIA data are described.

Today, the majority of quantitative proteomics takes the bottom-up approach. Proteins from a sample are digested and the resulting peptides are analyzed by LC-MS/MS, which infer the identities and quantities of the proteins. An alternative method is to separate and quantify proteins on a 2D gel of which selected proteins (for example, those showing significant changes in the experimental condition) are subsequently identified by protease digestion and DDA. This is described in Chapter 13.

There are several specific topics that are less studied but are becoming equally important in neuroscience, including the analysis of neuropeptidome (Chapter 14), the mass spectrometric imaging of brain sections (Chapter 15), and the quantitative proteomics analysis of primary neuronal culture (Chapter 16).

In this book we have covered most of the current neuroproteomics technologies. I believe that the various chapters will be of use to help the readers to design and carry out their own proteomics experiments in optimal ways.

References

1. Lepeta K, Lourenco MV, Schweitzer BC, Martino Adami PV, Banerjee P, Catuara-Solarz S, de La Fuente Revenga M, Guillem AM, Haidar M, Ijomone OM, Nadorp B, Qi L, Perera ND, Refsgaard LK, Reid KM, Sabbar M, Sahoo A, Schaefer N, Sheean RK, Suska A, Verma R, Vicidomini C, Wright D, Zhang XD, Seidenbecher C (2016) Synaptopathies: synaptic dysfunction in neurological disorders - a review from students to students. J Neurochem 138:785–805

2. Serrano-Pozo A, Frosch MP, Masliah E, Hyman BT (2011) Neuropathological alterations in Alzheimer disease. Cold Spring Harb Perspect Med 1:a006189

3. Koopmans F, Ho JTC, Smit AB, Li KW (2018) Comparative analyses of data independent acquisition mass spectrometric approaches: DIA, WiSIM-DIA, and untargeted DIA. Proteomics 18(1):1700304

4. Pandya NJ, Koopmans F, Slotman JA, Paliukhovich I, Houtsmuller AB, Smit AB, Li KW (2017) Correlation profiling of brain subcellular proteomes reveals co-assembly of synaptic proteins and subcellular distribution. Sci Rep 7:12107

5. Sialana FJ, Gulyassy P, Majek P, Sjostedt E, Kis V, Muller AC, Rudashevskaya EL, Mulder J, Bennett KL, Lubec G (2016) Mass spectrometric analysis of synaptosomal membrane preparations for the determination of brain receptors, transporters and channels. Proteomics 16:2911–2920

6. Counotte DS, Goriounova NA, Li KW, Loos M, van der Schors RC, Schetters D, Schoffelmeer AN, Smit AB, Mansvelder HD, Pattij T, Spijker S (2011) Lasting synaptic changes underlie attention deficits caused by nicotine exposure during adolescence. Nat Neurosci 14:417–419

7. Van den Oever MC, Goriounova NA, Li KW, Van der Schors RC, Binnekade R, Schoffelmeer AN, Mansvelder HD, Smit AB, Spijker S, De Vries TJ (2008) Prefrontal cortex AMPA receptor plasticity is crucial for cue-induced relapse to heroin-seeking. Nat Neurosci 11:1053–1058

8. Heo S, Diering GH, Na CH, Nirujogi RS, Bachman JL, Pandey A, Huganir RL (2018) Identification of long-lived synaptic proteins by proteomic analysis of synaptosome protein turnover. Proc Natl Acad Sci U S A 115:E3827–E3836

9. Umoh ME, Dammer EB, Dai J, Duong DM, Lah JJ, Levey AI, Gearing M, Glass JD, Seyfried NT (2018) A proteomic network approach across the ALS-FTD disease spectrum resolves clinical phenotypes and genetic vulnerability in human brain. EMBO Mol Med 10:48–62

10. Ping L, Duong DM, Yin L, Gearing M, Lah JJ, Levey AI, Seyfried NT (2018) Global quantitative analysis of the human brain proteome in Alzheimer's and Parkinson's disease. Sci Data 5:180036

11. Monti C, Colugnat I, Lopiano L, Chio A, Alberio T (2018) Network analysis identifies disease-specific pathways for Parkinson's Disease. Mol Neurobiol 55:370–381

12. Hondius DC, Eigenhuis KN, Morrema THJ, van der Schors RC, van Nierop P, Bugiani M, Li KW, Hoozemans JJM, Smit AB, Rozemuller AJM (2018) Proteomics analysis identifies new markers associated with capillary cerebral amyloid angiopathy in Alzheimer's disease. Acta Neuropathol Commun 6:46

13. Hondius DC, van Nierop P, Li KW, Hoozemans JJ, van der Schors RC, van Haastert ES, van der Vies SM, Rozemuller AJ, Smit AB (2016) Profiling the human hippocampal proteome at all pathologic stages of Alzheimer's disease. Alzheimers Dement 12:654–668

14. Suzuki T, Kametani K, Guo W, Li W (2018) Protein components of post-synaptic density lattice, a backbone structure for type I excitatory synapses. J Neurochem 144:390–407

15. Datta S, Malhotra L, Dickerson R, Chaffee S, Sen CK, Roy S (2015) Laser capture microdissection: Big data from small samples. Histol Histopathol 30:1255–1269

16. Drummond E, Nayak S, Faustin A, Pires G, Hickman RA, Askenazi M, Cohen M, Haldiman T, Kim C, Han X, Shao Y, Safar JG, Ueberheide B, Wisniewski T (2017) Proteomic differences in amyloid plaques in rapidly progressive and sporadic Alzheimer's disease. Acta Neuropathol 133:933–954

17. Wong TH, Chiu WZ, Breedveld GJ, Li KW, Verkerk AJ, Hondius D, Hukema RK, Seelaar H, Frick P, Severijnen LA, Lammers GJ, Lebbink JH, van Duinen SG, Kamphorst W, Rozemuller AJ, Netherlands Brain Bank, Bakker EB, International Parkinsonism Genetics Network, Neumann M, Willemsen R, Bonifati V, Smit AB, van Swieten J (2014) PRKAR1B mutation associated with a new neurodegenerative disorder with unique pathology. Brain 137:1361–1373

18. Chen N, van der Schors RC, Smit AB (2011) A 1D-PAGE/LC-ESI linear ion trap orbitrap MS approach for the analysis of synapse proteomes

and synaptic protein complexes. In: Li KW (ed) Neuroproteomics. Humana Press, Totowa, NJ, pp 159–167

19. Wisniewski JR, Zougman A, Nagaraj N, Mann M (2009) Universal sample preparation method for proteome analysis. Nat Methods 6:359–362

20. Hughes CS, Foehr S, Garfield DA, Furlong EE, Steinmetz LM, Krijgsveld J (2014) Ultrasensitive proteome analysis using paramagnetic bead technology. Mol Syst Biol 10:757

21. Chen N, Koopmans F, Gordon A, Paliukhovich I, Klaassen RV, van der Schors RC, Peles E, Verhage M, Smit AB, Li KW (2015) Interaction proteomics of canonical Caspr2 (CNTNAP2) reveals the presence of two Caspr2 isoforms with overlapping interactomes. Biochim Biophys Acta 1854:827–833

22. Li KW, Chen N, Klemmer P, Koopmans F, Karupothula R, Smit AB (2012) Identifying true protein complex constituents in interaction proteomics: the example of the DMXL2 protein complex. Proteomics 12:2428–2432

23. Pandya NJ, Klaassen RV, van der Schors RC, Slotman JA, Houtsmuller A, Smit AB, Li KW (2016) Group 1 metabotropic glutamate receptors 1 and 5 form a protein complex in mouse hippocampus and cortex. Proteomics 16:2698–2705

24. Chen N, Pandya NJ, Koopmans F, Castelo-Szekelv V, van der Schors RC, Smit AB, Li KW (2014) Interaction proteomics reveals brain region-specific AMPA receptor complexes. J Proteome Res 13:5695–5706

25. Liu F, Lossl P, Rabbitts BM, Balaban RS, Heck AJR (2018) The interactome of intact mitochondria by cross-linking mass spectrometry provides evidence for coexisting respiratory supercomplexes. Mol Cell Proteomics 17:216–232

26. Liu F, Rijkers DT, Post H, Heck AJ (2015) Proteome-wide profiling of protein assemblies by cross-linking mass spectrometry. Nat Methods 12:1179–1184

27. McAlister GC, Nusinow DP, Jedrychowski MP, Wuhr M, Huttlin EL, Erickson BK, Rad R, Haas W, Gygi SP (2014) MultiNotch MS3 enables accurate, sensitive, and multiplexed detection of differential expression across cancer cell line proteomes. Anal Chem 86:7150–7158

28. Ludwig C, Gillet L, Rosenberger G, Amon S, Collins BC, Aebersold R (2018) Data-independent acquisition-based SWATH-MS for quantitative proteomics: a tutorial. Mol Syst Biol 14:e8126

Chapter 2

Dissection of Rodent Brain Regions: Guided Free-Hand Slicing and Dissection of Frozen Tissue

Sabine Spijker, Leonidas Faliagkas, and Priyanka Rao-Ruiz

Abstract

Previously we reported on a free-hand dissection guide for proteomics samples, where the major focus was on dissection of fresh brain tissue. Although the fresh brain provides clear advantages, such as being able to see structures in very great detail by differences in color and shade, it also has some disadvantages. This particularly becomes evident when processing multiple samples within a short time-window, e.g., after a physiological stimulus. In this chapter, we discuss some of the current methods to take tissue samples from frozen tissue, keeping in mind the requirements for subcellular isolation (e.g., synaptosomes) of proteomics samples. In addition, we provide a step-by-step protocol to improve standardization of tissue dissection in your lab.

Key words Guided free-hand dissection, Mouse, Rat, Brain, Frozen tissue

1 Introduction

Often, samples for various –omics techniques are harvested from frozen brain tissue. Specifically, smaller brain regions, e.g., amygdala, can be dissected or punched out in a relatively precise manner when using specific anatomical landmarks. As studying gene and protein expression after a physiological stimulus becomes more popular, the processing of these brain samples becomes a limiting factor. Hence, when timing is important and the total sample size (n-number, different genotypes or treatment groups) increases, it is easier to first freeze these brain samples and only later perform the dissections.

What makes sample preparation for proteomic special is that when one wants to obtain subcellular fractions, specific requirements come into place, in terms of maximal thickness of slices (300–400 μm). As such, the techniques described here, as well as the specific guided free-hand dissection method, focus on choices to be made on how to freeze the brain (see Subheading 1.1) and how to slice the brain (see Subheading 1.2). The various steps in

Ka Wan Li (ed.), *Neuroproteomics*, Neuromethods, vol. 146,
https://doi.org/10.1007/978-1-4939-9662-9_2, © Springer Science+Business Media, LLC, part of Springer Nature 2019

the protocol (*see* Subheadings 2 and 3) provide detailed photographs alongside their corresponding anatomical pictures from the Allen brain atlas. It is important to note that although we detail our procedures for mouse tissue, a similar strategy can be implemented for any other species, provided that specific reference Atlases are available that are checked for strain, sex, and age.

1.1 What to Dissect and How to Freeze the Brain

As discussed before [1], dissecting regions of interest from frozen tissue has the advantage that more regions are accessible at the same time than with fresh tissue. Fresh dissection is less precise, and therefore should be focused on regions that are either hard to dissect in the frozen brain or that are relatively large, such as the dorsal striatum, thalamus, and the entire prefrontal cortex. In principle, with frozen dissection, various subregions can be harvested in parallel by precisely cutting out the region of interest from a brain slice, e.g., the prelimbic and infralimbic prefrontal cortex, and nucleus accumbens (ventral striatum).

Frozen tissue is especially suitable for tissue punching of small regions like the amygdala or ventral tegmental area (VTA). Note however, that specific anatomical features that are well distinguishable in the fresh brain might be harder to see in frozen sections (Fig. 1). As such, clear anatomical hallmarks must be used to ensure precise dissection. In this protocol we will discuss several steps to decrease this possible source of variation.

Prior to material collection, it is wise to consider the type of freezing protocol to be utilized and the orientation in which the brain is frozen. With regards to freezing protocols, a routine technique involves snap-freezing tissue into relatively inert cryogenic liquids, such as liquid nitrogen (−196 °C), or in 2-methylbutane

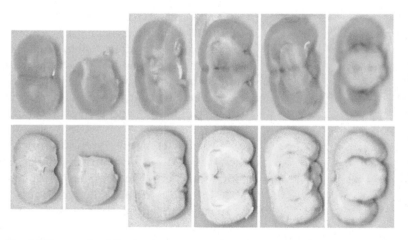

Fig. 1 Difference in ability to observe anatomical landmarks in frozen and "fresh" mouse brain sections. The same frozen sections (lower) were thawed (upper) in order to view the distinction in color and anatomical landmarks. The thawed version was covered in Aquatex in order to prevent drying during photography

(formerly called isopentane) that is precooled (−78 °C) in dry-ice sludge. The advantage of snap-freezing protocols is that it keeps degradation of proteins or RNA to a minimum, while preserving the structure well. However, there are a few disadvantages attached to these methods in relation to subsequent slicing for tissue dissection; the entire brain is frozen in the form it was just prior to freezing, hence special care should be taken to maintain anatomical integrity. Another disadvantage is that due to the unequal speed of freezing, friction within the brain might lead to tissue cracks. This typically occurs with larger samples like the rat brain, but may also happen with the mouse brain. When solely used for dissection this does not have to be a problem, although specific anatomical landmarks might be harder to see. Furthermore, both cryogenic liquids are highly hazardous (Box 1); liquid nitrogen due to its low temperature and liquid-to-gas expansion ratio and 2-methylbutane due to its extremely high toxicity and burden on the environment. Lastly, when used for free-hand slicing, as we discuss here, the natural bulged forms of the brain will make it hard to keep it in a steady position during slicing on a flat surface/carrier. Therefore, we will discuss a fast-freezing option using dry-ice (Box 1) only. Note, however, that slower freezing, than that used with cryogenic liquids, could result in the formation of ice-crystals within the brain that may severely damage its (ultra)structure. This is of importance when the brain specimen is used for both proteomics sampling and for immunological staining of tissue sections. For details see Cunningham & Doran 2012 (https://www.leicabiosystems.com/pathologyleaders/freezing-biological-samples/).

Box 1: Cryogenic Liquids and Solids for Brain Freezing

Liquid nitrogen:

Has a high liquid-to-gas expansion ratio at room temperature. Therefore, breaking a bucket of liquid nitrogen in a small room can yield a high pressure. Moreover, it can then be an asphyxiant. Watch out for cold-burns when spilling droplets of liquid on unprotected skin.

2-Methylbutane/Methylbutane/Isopentane:

Is extremely volatile and flammable (GHS category H224) at room temperature. Best to store cooled, especially when using it as cryogenic compound. Toxicity for aspiration (H304, H336) and the burden on the aquatic environment (H411) form major disadvantages of its use, and urge for alternatives.

Dry ice (solid CO2):

Sublimates into CO_2 gas, leaving no residual. Watch out for cold-burns when touching the pellets.

Fig. 2 Fast-freezing the mouse brain on dry ice. The brain is put in the desired position while on normal ice (**a**) and frozen on dry ice (**b**) while being on the same carrier. The aluminum foil prevents the brain from sticking to the carrier and can be used to immediately wrap the brain for storage purposes. This procedure keeps the frozen brain in a steady position for subsequent (guided) free-hand slicing (**c**)

The dry-ice freezing option uses a limited set of materials (*see* below) and typically takes ~1–2 min longer than the snap-freezing method with size of tissue influencing freezing time. For this procedure, the brain is taken from the skull and put on a piece of aluminum foil on a metal carrier in the desired orientation/position, while being cooled on normal ice. Subsequently the foil and carrier are placed on dry ice, allowing the brain to freeze (Fig. 2a, b), where the speed of freezing can be adjusted by placing dry ice parts or pellets near the brain. As such, the brain will freeze and be maintained in a steady position (Fig. 2c) for subsequent free-hand slicing. It is, however, important to keep in mind that the way of positioning the brain (ventral part up, or dorsal part up) influences the regions to be dissected. For example, when the region of interest is ventrally located, it is wiser to freeze the brain with the ventral surface facing up because (1) contact with the metal carrier deforms normal anatomy and (2) with freehand slicing standardized vertical cuts are often hard to maintain and so linear and volumetric variations in the plane of slicing tend to be the smallest closer to the surface from where slices are made.

A last important step is storage of the collected brain samples. Mostly, bulky samples like entire brains are stored at −80 °C, where it is essential to protect the sample from freeze-drying, a process that occurs at low temperatures over time. Therefore, every brain should be wrapped in a piece of aluminum foil. Either write the animal code on the foil and/or add a pencil- or laser-printed paper slip to it for long-term storage and retrieval. Store all brain specimens together in a carton/plastic box (−80 °C) and add a pencil- or laser-printed experiment description.

1.2 How to Slice the Brain?

The most important choice to make is the method used to slice the brain. In principle, there are four options detailed in Box 2. Each option has advantages and disadvantages, but due to its relative simplicity, low costs but good possibility to standardize, we will discuss option 4; using a self-constructed device for guided free-hand slicing.

Box 2: Slicing Options for Frozen Brain

Cryostat:

Check your departments' cryostat for the option to make thick (>200 μm) sections. Older machines can be 'preloaded' for example by turning the wheel 3× without cutting and reversing it and making the actual cut at the 4th turn (set at 50 μm). Slice at −12 to −10 °C. Check whether this option is suitable for your samples with a few test brains first.

Advantage: Very precise slicing. Ideal for relatively thin slices (<100 μm). Works better for rat than for mouse brains due to size.

Disadvantage: Time-consuming, for thicker slices of a mouse brain (>100–200 μm), it is essential to embed the brain in OCT compound's like Tissue-Tek®, to prevent curling and crumbling of brain slices. Thin slices are harder to process for subcellular fractionation in proteomics.

For the options below, one needs a cooling device that is ideally set at −12 to −10 °C. In principle this is possible in a cryostat. Note however that modern cryostats have specific temperature controls for the tissue specimen stage, whereas the chamber itself is at −20 °C, which is too cold. Also keep in mind that you will need a metal non-iron plate for this, in order to not damage the cryostat table. Always check whether this option is suitable for your samples with a few test brains first. It is not possible to perform simultaneous multiple slicing as the low temperature will deform the entire brain by the pressure put onto it. Lowering the temperature (−15 °C) results in more difficulty to slice and hence more pressure needed. The protocol described in this chapter in more detail uses the guided free-hand slicing and dissection.

Brain slicer matrix:

Best to combine with a cryogenic liquid freezing protocol, due to the curvature of the matrix.

Advantage: Relatively precise slicing, thinner than the free-hand options. Different options are available for different sized brains. Relatively standardized slicing.

Disadvantage: Expensive device. Difficult to hold the brain in the same position while cutting.

Freehand slicing:

Combine with dry-ice freezing protocol.

Advantage: Cheapest and easiest option. Ideal when dissecting large tissue chunks.

Disadvantage: Not precise. Large variation from sample-to-sample and between people.

Guided free-hand slicing:

Combine with dry-ice freezing protocol. Self-make your own slice-guide.

(continued)

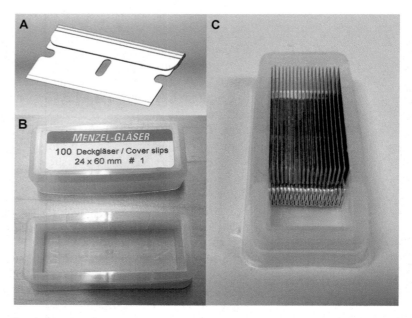

Fig. 3 Free-hand slicing guide. (**a**, **b**) Standard single-edge razor blade (e.g., Fisherbrand 12–640; **a**) was pasted into a microscopy cover slip box (**b**). (**c**) This specific guide contains 19 blades with an inter-blade distance of ~1 mm

Advantage: Cheap and easy construction for cutting guide. Flexible in size. Easy to use. Improved precision over free-hand slicing and hence provides medium-high standardization of sample collection.

Disadvantage: Cutting distance is determined by the width of the single edge blade (Fig. 4). Yet creative minds might see a way around this limitation.

This device (Fig. 3) will be used to make precuts in the frozen brain to guide subsequent free-hand slicing (Fig. 4). For standardization purposes, it is important to achieve a constant inter-blade distance. In this case, this was achieved by pasting 19 standard single-edge razor blades into a cover slip box (Fig. 3), creating a distance of ~1 mm, where the number of blades depends on the size of the lid you put the blades in. In our case, the rims on the outside of the box enable easy handling of the device.

It is important to adjust the type of slicing and the standardization to your needs. In Fig. 4, we have indicated to have one precut at a fixed position, i.e., at the most posterior contact side of left and right cerebrum (Fig. 4c) or at the border of the most anterior side of the hypothalamus at the side of the optic chiasm (Fig. 4d).

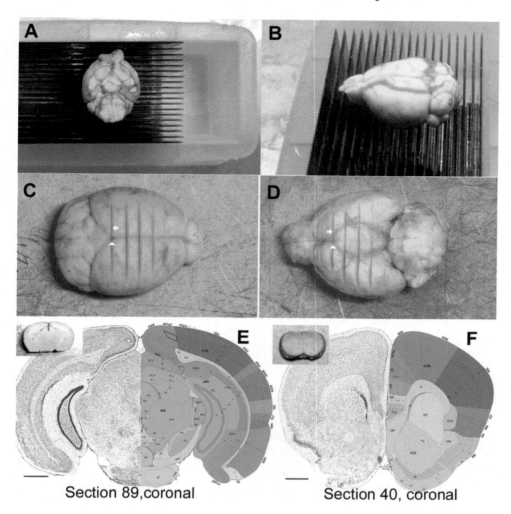

Fig. 4 Free-hand slicing guide in action. (**a–d**) Brains for dorsal dissection (**a, b**) or ventral dissection (**b, d**) were precut (**c, d**, respectively) at a fixed anatomical position (arrowhead). Note the visible contact site of freezing and subsequent deformation at the ventral and dorsal site in panels **a** and **b**, respectively. Subsequently, using a separate blade, free-hand coronal slices were generated based on the precuts, while keeping the brain on a metal non-iron plate in a cryostat (−10 °C). (**e, f**) The corresponding Allen Brain reference atlas picture of the slice (arrows indicate direction of view) indicated in panels **c, d** is shown, with the actual slice as inset

2 Materials

The products used are listed below. Comparable products from other suppliers should also be effective. Equipment visible on the photographical and schematic representation of the dissections (Figs. 3–6, 9) is indicated by *.

2.1 *Brain Removal*

See description given in Spijker 2011 [1].

2.2 Brain Freezing

1. Favorably 2–3 metal iron-free plates [1]. Alternatively, take a solid plastic item and wrap aluminum foil firmly around it. Always keep a separate piece of foil to put the brain on.

2. Aluminum foil in pieces of ~6 × 6 cm (mouse) or 10 × 10 cm (rat).

3. Forceps as indicated previously [1].

4. Dry ice pellets in a polystyrene box with lid.

2.3 Brain Dissection

1. Favorably 1 metal iron-free plate [1]. Alternatively, take a solid plastic item, and wrap aluminum foil firmly around it. Due to brain slicing this latter material is not preferred.

2. Cotton gloves (e.g., Cara cotton gloves (medium, Walmart #551140545)) to be worn under normal Nitrile or Latex gloves to reduce heat transfer and to stay comfortable while dissecting in a cryostat.

3. Razor blades (e.g., Fisherbrand 12–640).

4. Scalpel handle (FST # 10003–12 Scalpel Handle #3—12 cm) with scalpel (FST # 10011–00), as illustrated in Spijker 2011,

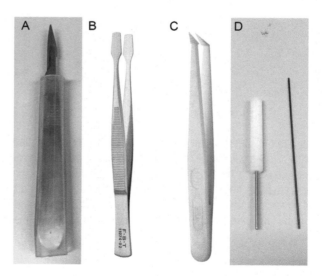

Fig. 5 Specific dissection tools. (**a**) Scalpel handle (FST # 10003–12 Scalpel Handle #3—12 cm) with scalpel (FST # 10011–00), where the handle is covered with clear PVC tubing to reduce heat transfer while dissecting. (**b, c**) Forceps to keep the brain in a steady position while free-hand slicing; straight cover glass forceps (FST11074–02, **b**). To reduce heat transfer while dissecting the brain, consider the plastic (Acetyl) version (FST11700–02; **c**). Although it has a smaller surface, this 45° angled version has a pointed tip that eases picking up the dissected brain areas. (**d**) Tissue puncher (FST # 18036–19) and metal expeller

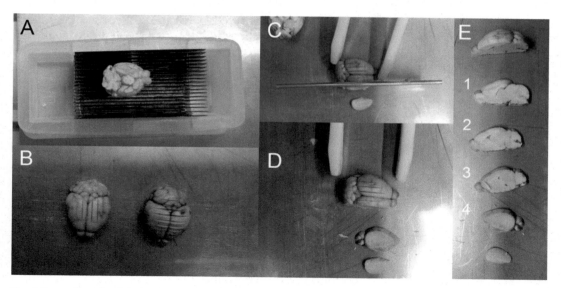

Fig. 6 Generation of a digitally archived sagittal series. (**a, b**) The brain (frozen with ventral side down) is precut with the slice-guide (**a, b**). (**c, d**) Subsequently, while holding the brain with a plastic forceps, sagittal sections are made using a razor blade (**c, d**). (**e**) The consecutive sections are numbered (1–4) and are detailed upon in Fig. 7

Fig. 2c [1]. A plastic handle (FST # 10000–11) of the disposable scalpels reduces heat transfer to the brain sample during dissection, and hence might be the preferred choice.

5. Optional: Use clear PVC (Vinyl) tubing to cover the stainless-steel handle of FST # 10003–12 and hence have a cheap and endurable solution to reduce heat transfer (Fig. 5a).

6. Straight forceps with larger surface area to firmly hold the brain without damaging it. The stainless steel version FST11074–02 (straight cover glass forceps; Fig. 5b) would suffice, but very suitable plastic versions (e.g., FST11700–02; Figs. 5c and Fig. 6) are available as well.

7. Tissue puncher (FST # 18036–19 for 1 mm ø), if applicable for the type and size of tissue (e.g., amygdala, VTA). Note that it creates a bit of damage (~0.2–0.6 mm) to the surrounding tissue,

8. Create the guided dissection tool with the following (*see* Subheading 1.2 and Fig. 3):

    • Microscopic coverslip box. Menzel-Gläser, 24 × 60, #1 (e.g., VWR 631–0973; Fig. 3b) ∗. Alternative: use the 25 × 50, #1 (e.g., VWR 631–0854).

    • Standard razor blades (Fisherbrand 12–640).

    • Glue.

9. Atlases to compare your slices with; e.g., mouse brain digital ones from Allen Brain Atlas (http://mouse.brain-map.org/static/atlas) or from Matt Gaidica based on Paxinos and Franklin 2001 [2] (http://labs.gaidi.ca/mouse-brain-atlas), as well as the rat version based on Paxinos and Watson 2006 [3] (http://labs.gaidi.ca/rat-brain-atlas). The Allen Brain atlas pictures have the advantage that specific subregions can be highlighted and are easy to find. The Paxinos picture indicates both the distance relative to Bregma and the Interaural point, as well as the medial-lateral distance (in mm). Both mouse atlases are based on a P56 mouse brain, hence variation in the exact coordinates may occur when using older animals.

3   Methods

3.1   Brain Removal

See description given in Spijker 2011 [1].

3.2   Brain Freezing

1. Prepare (incl. Marking for specimen) aluminum foil pieces to wrap the frozen brain, as well as preparing paper slip (*see* Subheading 1.1) to mark the brain specimen (genotype, treatment, experiment, animal number).

2. Place the aluminum foil on the metal iron-free plate that is on ice.

3. Transfer the brain after removal from the skull onto the aluminum foil. Place with the dorsal side up when the area of interest is on the dorsal side.

4. Transfer the entire plate onto the dry ice (pellets) to freeze slowly. One can increase freezing speed by adding pellets onto the plate in close vicinity (but not touching) the brain.

5. Depending on the species (mouse ~2–4 min; rat ~3–6 min) fold the aluminum foil (with paper slip) to show the experimental marks (e.g., animal number) visible to ease later retrieval of a specific specimen.

3.3   Brain Dissection

It is advisable to practice the use of dissection several times. When sufficient reproducibility is achieved for brains of a specific strain, genotype, sex, and age, it is advisable to generate a standard brain series (Figs. 6 and 7) where the anterior and posterior sides of each slice is archived digitally and compared with available brain atlases. For such a series, one could make the frozen and the thawed version, as each provides clear and valuable anatomical hallmarks (*see* Fig. 7).

1. Transfer the aluminum foil-wrapped brains on dry ice to the cryostat.

2. Put the temperature of the cryostat at −10 °C (or the preferred slicing temperature).

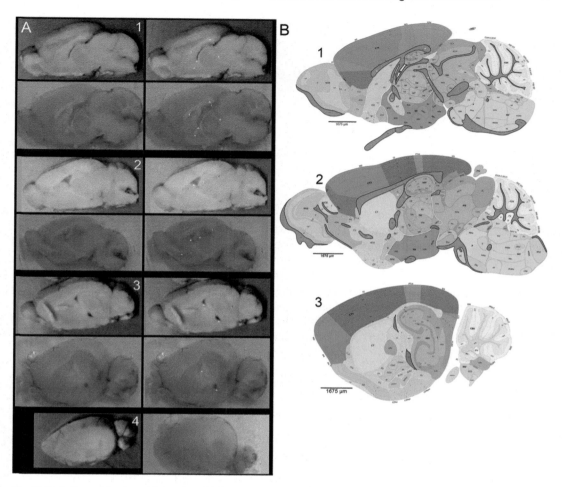

Fig. 7 Digitally archived sagittal series. (**a**) Frozen sections labeled 1–4, see Fig. 6, are depicted, as well as the thawed version covered with Aquatex. All pictures are duplicated to show specific anatomical landmarks on the right side. **a1**: Third ventricle (white arrowhead), various white matter tracts (yellow arrowheads), and hippocampus (red arrowhead with yellow line) are indicated. Note that specific features are also visible on section 19 of the Allen Brain atlas (not shown here). **a2**: Lateral ventricle (white arrowhead), the fimbria (yellow arrowhead), the anterior commissure (turquois arrowhead), and hippocampus (red arrowhead with yellow line). **a3**: Dorsal and ventral part of the lateral ventricle (white arrowheads) between the white matter of, e.g., fimbria and internal capsule visible (not indicated), and hippocampus (red arrowhead with yellow line). Note that specific features (e.g., the shape of the hippocampus) are better visible on section 5 of the Allen Brain atlas (not shown here). (**b**) The corresponding Allen Brain reference atlas picture of slices 1–3 (section 18, 14, and 6, respectively) are shown, as slice 4 falls outside the range depicted in the Allen Brain Atlas. In **b1** and **b2** the white matter tracts/fibers are highlighted in purple, and in **b1** the third ventricle (V3) is highlighted in white. In **b3** the lateral ventricle is highlighted in purple. Note. Arrowheads as depicted in (**a**) are visible at the corresponding spot in (**b**)

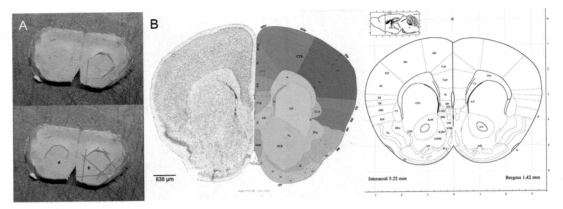

Fig. 8 Dissection of dorsal and ventral striatum. (**a**) Brain slice similar to the one shown in Fig. 4d, f. The lower picture shows the same slice but now with marker lines indicating the border of the dorsal striatum and fmi (green), the anterior commissure (aca; blue), and the incisions made to obtain the dorsal (caudate putamen, CP(u)) and ventral (nucleus accumbens (ACB/Acb), core and shell) part of the striatum. Incisions were made using a scalpel and knife (FST # 10011–00, see Fig. 5a). Note that depending on the thickness of the slice and the anterior–posterior differences in size of the brain area, these incisions can be made on either side, where cutting in an angle could help to adjust for these differences. (**b**) Digital atlas pictures (*left, Allen Brain section 40; lower, Paxinos figure 19 by Matt Gaidica*) corresponding to the anterior side of the slice

3. Place a metal non-iron plate on the cryostat table. After 2 min place the brains on the plate, and let them acclimatize for 2–3 h.

4. Place all dissection tools in the cryostat ~5 min before the start.

5. Prepare Eppendorf vials to hold the dissected brain areas. Label all tubes well, both on the top and the side. Note that it might be handy to dissect more areas from the same slice, even if the tissue is not used immediately. Especially for large experiments, it is ethically preferable to use these additional brain areas for a pilot study at a later moment in time, rather than to restart a new experiment with new animals.

6. Unwrap the first brain and use the dissection guide to make incisions on the dorsal or ventral part (*see* Fig. 4).

7. Dissect (Fig. 8) or punch out the region of interest (Fig. 9).

8. Gather the tissue of the region of interest in Eppendorf tubes, store at −80 °C until further processing.

9. After the preferred brain areas have been taken out, put back the remainder of the brain (if applicable) in the aluminum foil and store at −80 °C for later use.

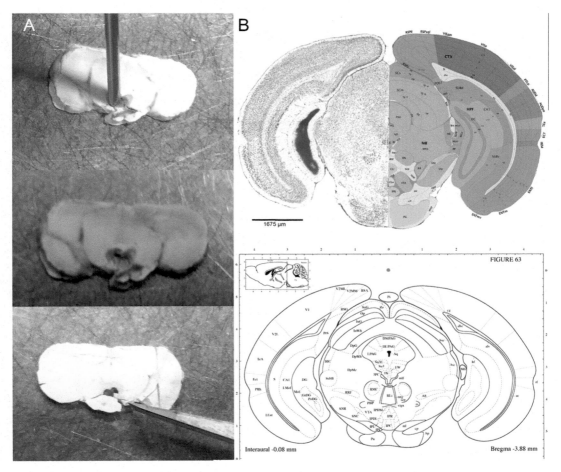

Fig. 9 Punching out VTA-enriched mid-brain. (**a**) The posterior side from the last coronal section was used to punch out VTA-enriched mid-brain (*upper two pictures*) into the anterior direction using a 1 mm ø tissue puncher (FST # 18036–19). On the anterior side (*bottom picture*), the punch is visible as well. Note that a start was made in dissecting the ventral subiculum on the right side. (**b**) Digital atlas pictures (*upper, Allen Brain section 90; lower, Paxinos figure 63 by Matt Gaidica*) corresponding to the posterior side of the slice

References

1. Spiker S (2011) Dissection of rodent brain regions. In: Li KW (ed) Neuroproteomics. Humana Press, Totowa, NJ, pp 13–26
2. Paxinos G, Franklin KBJ (2001) The mouse brain in stereotaxic coordinates: hard cover edition. Elsevier
3. Paxinos G, Watson C (2006) The rat brain in stereotaxic coordinates: hard cover edition. Elsevier

Chapter 3

Isolation of Synapse Sub-Domains by Subcellular Fractionation Using Sucrose Density Gradient Centrifugation: Purification of the Synaptosome, Synaptic Plasma Membrane, Postsynaptic Density, Synaptic Membrane Raft, and Postsynaptic Density Lattice

Tatsuo Suzuki, Yoshinori Shirai, and Weidong Li

Abstract

A protocol presents a purification of postsynaptic density (PSD), from rat brain by subcellular fractionation using solubilization of membrane with Triton X-100 and sucrose density centrifugation. The protocol also includes purification of other synapse sub-domains such as synaptosome, synaptic plasma membrane (SPM), synaptic membrane raft, PSD lattice, P_1 (nuclei and cell debris), P_2 (crude mitochondria fraction), S_3 (soluble fraction), and P_3 (microsomal fraction). The PSD purification method presented in this text is the one established by Siekevitz group. The PSDs obtained by this method are mainly excitatory type I PSDs. These methods are useful for biochemical analyses such as identification of proteins associated with these sub-domains by proteomics methods and western blotting, and morphological analyses at the electron microscopic level. The purification protocol for the synaptic membrane raft using sucrose gradient ultracentrifugation is a useful means by which to analyze the relationship between the PSD and synaptic membrane raft by isolating both preparations simultaneously.

Key words Synaptosome, Synaptic plasma membrane, Postsynaptic density, Postsynaptic membrane raft, PSD lattice, Subcellular fractionation, Detergent-insoluble cytoskeleton, Detergent-insoluble membrane

1 Introduction

Isolation of subcellular compartment is a useful approach to analyze the subcellular complexes at the molecular level. We describe methods to isolate synapse sub-domains including P_1 (nuclei and cell debris), P_2 (crude mitochondria fraction), synaptosome, synaptic plasma membrane (SPM), postsynaptic density (PSD), synaptic membrane raft, and PSD lattice by subcellular fractionation using density gradient centrifugation. History of development of method for isolation of synaptic complex and PSD is concisely summarized

Ka Wan Li (ed.), *Neuroproteomics*, Neuromethods, vol. 146,
https://doi.org/10.1007/978-1-4939-9662-9_3, © Springer Science+Business Media, LLC, part of Springer Nature 2019

previously [1]. The method for PSD purification established by Siekevitz's laboratory [2–4] has been widely used. The methods introduced in this text are basically the same as those used in his laboratory. Both short and long procedures are stated. The method is applicable to brains from at least dog, rat, mouse, and human [2, 5, 6], different brain regions [4, 7–10] and brains in various developmental stages [11]. Protein yield of synaptosome, SPM, and PSD (short and long procedures, respectively) are approximately 14.7, 6.2, and 0.26 and 0.1 mg per 1 g original forebrain of adult rat, respectively, but may fluctuate by unknown factor(s). Percentages of PSD protein per amount of total protein in the forebrain, synaptosome, and SPM are summarized in Table 1. Protein yields of these fractions change depending on the age of the animals used [11]. The short procedure, in which PSD is purified from Triton X-100 (TX-100)-treated synaptosomes, has been widely used and is now a standard method. Protein profiles of the PSD isolated by short and long procedures in one-dimensional gel are similar but not identical (Fig. 1) [2, 12]. Contents of neurofilament proteins, at least partly contaminants [12], are higher in the PSD prepared by short procedure [12]. Protein yield is also different [12]. Major constituent proteins are the same between the two preparations, while the mass spectrometric (MS) analysis revealed that only 60–75% proteins in these two PSD fractions are common (T. Suzuki, unpublished data). Purified PSD fraction also contains mRNAs encoding various kinds of proteins [13].

In the early methods to prepare synaptic junctional complex and PSD, p-iodonitrotetrazorium violet (INT) was used to separate mitochondria by producing heavy formazan in mitochondria [14–16]. However, it was found that INT causes undesirable oxidation of proteins and artificially cross-links synaptic proteins [16–19]. Structures of the isolated PSD are tightened by disulfide bonds formed during the PSD isolation using INT. It is suggested that the artificial disulfide bonding of PSD proteins during isolation may occur even in the absence of INT [1, 20, 21]. Artificial cross-linking of postsynaptic proteins during isolation gives resistance of the isolated PSD to various treatments including detergent solubilization [1, 20, 21]. Blocking of disulfide formation is required for preparing PSD for analyses of its structural and physiological properties.

It is desirable to prepare synaptic sub-compartments from freshly dissected brains. PSD fraction can also be prepared from frozen brains [3], which is convenient, in particular, when purifying it from human specimens. However, special attention should be paid when collecting brain tissue, because some proteins, in particular Ca^{2+}/calmodulin-dependent protein kinase II (CaMKII), accumulate to PSD in a short duration after decapitation [22]. Accumulation of CaMKII is accelerated at room temperature or 37 °C. Tubulin also accumulates to PSD fraction in a relatively

Table 1
Summary of PSD protein content

Subfractions	Protein recovery mg/g FB (mean ± S.D.)	(%)	(%)	(%)
Total protein	92.6 ± 21.1 ($n = 3$)	100		
Synaptosome	14.7 ± 4.2 ($n = 5$)	15.9	100	
SPM	6.24 ± 1.77 ($n = 4$)	6.74	42.4	100
PSD	0.263 ± 0.109 ($n = 11$)	0.284	1.79	4.21

FB forebrain
PSD was purified from frozen forebrains of rat (6 weeks old, male) by short protocol
Values were calculated from the protein yields for total homogenate, synaptosome, SPM, and PSD prepared from adult Wistar rat forebrain

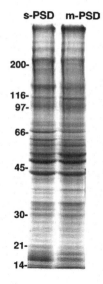

Fig. 1 Protein profiles of PSD fractions purified by short and long procedures. PSD fractions was purified from rat forebrains (Wistar male, 6 weeks old) and separated by 7–17% gradient polyacrylamide gel. s-PSD and m-PSD refer to PSD fractions prepared by short and long procedures via TX-100 treatment of synaptosome and SPM, respectively. Molecular weights are shown in kDa on the left

long time period at 4 °C after decapitation [23]. Attention should also be paid to "cold-induced exodus of postsynaptic proteins" [24]. Exposure of neuron to coldness causes rapid disassembly of unstable microtubules that are present in the spine and associated with PSD. Various proteins also exit from PSD, and spine morphology, at least some, may change by this microtubule disassembly.

The method stated in this text is useful to prepare the fraction enriched in the PSDs of asymmetric type I excitatory synapses, but not of the inhibitory neurons, such as those in the cerebellum. Protein yield of cerebellar PSD fluctuates; sometimes very low due to unknown reasons. Preparation of type II inhibitory PSD has been reported [25]. Method for the purification of PSD using sonication but not detergent has also been reported [26], but up to now the method has been reported only once to the best of author's knowledge.

"One-Triton" PSD and "Two-Triton" PSD are prepared as a pellet after centrifugation of TX-100-treated synaptosomes [27–29]. "One-Triton" PSD contains detergent-resistant membrane (DRM) with light buoyant density, which is also TX-100 insoluble at 4 °C and floats on the 1.0 M sucrose layer [25, 30]. Recently, it is demonstrated that "One Triton PSD" also contains type II GABAergic inhibitory PSD [25].

Nonionic detergent TX-100 is usually used to purify PSD fraction. High-quality TX-100 should be used. Other detergents, such as deoxycholate (DOC) [31, 32], n-octyl β-D-glucoside (OG) [1, 33], and N-lauroyl sarcosinate (NLS) [16, 27, 32], have also been used. NLS, a strong ionic detergent, nearly completely solubilizes PSD components when oxidation is prevented with 1 mM N-ethylmaleimide (NEM) during isolation of PSD [1]. DOC-insoluble PSD shows clearly a lattice-like core PSD structure [32, 34], which is broken after NLS treatment [32]. OG is effective to solubilize rapidly the whole membrane [35, 36] and generally does not affect protein–protein interactions.

Presynaptic structure is unstable in alkaline solution, while postsynaptic structures are resistant [37]. Therefore, the synaptic junctional structures composed of both pre- and postsynaptic cytoskeletal structures can be prepared when synaptosome is solubilized with TX-100 at slightly acidic conditions [37].

Membrane rafts are distributed in both pre- and postsynaptic sites in all neuronal components, including axons, dendrites, and somas, in both immature and mature neurons. We term detergent-insoluble materials with light buoyant density purified from SPM at a low temperature the synaptic membrane rafts. The postsynaptic membrane raft (PSR) and PSD are two major postsynaptic signaling domains that interact physiologically, and it is thought that PSRs may be indispensable to PSD function [38]. PSRs may be essential components of the postsynaptic signaling machinery connected to PSDs, providing membrane anchor sites for PSD cytoskeletons/scaffolds, as well as signaling platforms and sites for membrane fusion and vesicular trafficking. Thus, it is likely that postsynaptic activities require both PSDs and PSRs. PSRs may play a role in synaptogenesis, growth, and maturation of developing PSDs, and support and regulate functions and plasticity of mature

PSDs. The notion of lipid/membrane rafts [39] was first proposed in 1997 [40]. PSRs and PSDs can be separated by density in vitro, though they are both detergent-insoluble at low temperatures and interact with one another [41–43].

In a recent concept, PSDs can be divided into two areas: the "PSD core" and the "PSD pallium," typically located at depths of 30–50 nm from the postsynaptic membrane and further 50–60 nm towards the cytoplasmic side, respectively [44] (Fig. 2). Purified PSD contains not only a PSD core but also a PSD pallium.

A network structure called the "junctional lattice" or the "PSD lattice" was identified long time ago by the extraction of the SPM, synaptic junction, or TX-100-insoluble PSD using the relatively strong detergent DOC (Fig. 3), and was proposed to be an under-

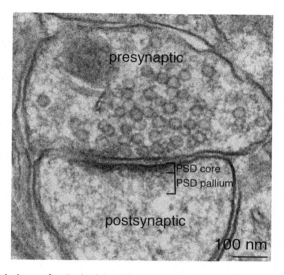

Fig. 2 Morphology of a typical type I excitatory synapse. The "PSD core" and "PSD pallium" regions are indicated. The image is from Dosemeci et al. [44]

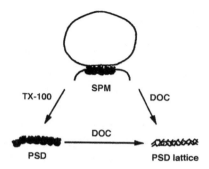

Fig. 3 The PSD lattice is visualized after treatment of the synaptic subfraction with DOC. The "PSD lattice" was identified in the 1970s by extraction of the SPM, synaptic junction, or TX-100-insoluble PSD using the relatively strong detergent DOC. The image is from Matus and Taff-Jones [32] with permission

lying structure of the PSD [31, 32, 34, 45]. However, until recently, the key components and molecular organization of the PSD lattice have not been determined [31, 32, 46].

2   Materials

Use distilled, double-distilled, distilled-and-deionized, or equivalent grade water. Using ultrapure water sometimes results in low protein yield of PSD (*see* **Note 1**). All chemicals should be of reagent grade. All stocks and working solutions are kept at −20 to −30 °C between uses to prevent bacterial and fungal growth. Make sure to mix up the solutions homogeneously after defreezing them, especially those containing dense sucrose solutions. All solutions should be kept at 4 °C or on ice during the subfractionation. TX-100 is susceptible to autoxidation upon exposure to air (*see* **Note 2**). Store the unused solution sealed and also avoid storage in direct light.

PSD material is extremely sticky to glass and cellulose nitrate and tend to aggregate very easily [2]. Therefore, usage of plastic (polyallomer) tubes and pipettes, in particular, after TX-100 treatment, is necessary to avoid undesirable absorption of PSD proteins to glasses.

Add protease inhibitors, phosphatase inhibitors, oxidization inhibitors, or RNase inhibitors as required. Addition of protease inhibitors results in increased yield of PSD proteins. It is desirable to purify PSD in the presence of iodoacetamide (IAA) or *N*-ethylmaleimide (NEM), which prevents harmful oxidation during the purification [1, 20, 21]. PSDs prepared in the presence of IAA (2 mM in solution A and solution B) are different from those prepared in the absence of IAA in detergent solubility, aggregation state of PSD, and possibly dynamic properties of PSD. Addition of dithiothreitol interferes with endogenous disulfide bondages necessary for the formation of normal PSD configuration [20, 21].

2.1   Preparation of P_1, P_2, Synaptosome, and PSD Fraction (Short Procedure)

1. 1 M $MgCl_2$ stock. Dissolve 20.33 g of $MgCl_2 \cdot 6H_2O$ (MW, 203.30) in 100 mL H_2O.

2. 1 M $CaCl_2$ stock. Dissolve 14.70 g of $CaCl_2 \cdot 2H_2O$ (MW, 147.02) in 100 mL H_2O.

3. 100 mM $NaHCO_3$ stock. Dissolve 1.68 g of $NaHCO_3$ in H_2O and make up to 200 mL with H_2O.

4. 1 M Tris–HCl (pH 8.1) stock. Dissolve 24.2 g of Tris(hydroxymethyl)aminomethane in H_2O (~150 mL) and adjust pH to 8.1 by HCl, and make up to 200 mL with H_2O.

5. 0.5 M HEPES/KOH (pH 7.4) stock.
   Dissolve 11.8 g of HEPES, adjust pH to 7.4 with KOH solution and make up to100 mL with H_2O.

6. Solution A (0.32 M sucrose, 1 mM $MgCl_2$, 0.5 mM $CaCl_2$, 1 mM $NaHCO_3$). Dissolve 109.6 g of sucrose in H_2O. Add 10 mL of 100 mM $NaHCO_3$, 1 mL of 1 M $MgCl_2$ and 0.5 mL of 1 M $CaCl_2$. Make up to 1000 mL with H_2O.

7. Solution B (0.32 M sucrose, 1 mM $NaHCO_3$). Dissolve 109.6 g of sucrose in H_2O. Add 10 mL of 100 mM $NaHCO_3$. Make up to 1000 mL with H_2O.

8. 1% TX-100, 0.32 M sucrose, 12 mM Tris–HCl (pH 8.1). Dissolve 109.6 g of sucrose in H_2O. Add 10 g of TX-100 (Sigma) and 12 mL of 1 M Tris–HCl (pH 8.1). Make up to 1000 mL with H_2O.

9. 1% TX-100, 150 mM KCl. Dissolve 2 g of TX-100 and 2.26 g of KCl in H_2O. Make up to 200 mL with H_2O.

10. 10 mM HEPES/KOH (pH 7.4)-40% glycerol. Dilute 4 mL of 0.5 M HEPES/KOH (pH 7.4) in H_2O (~80 mL). Add 80 g of glycerol and make up to 200 mL with H_2O.

11. Sucrose solution (1.0, 1.4, 1.5 and 2.1 M). Dissolve sucrose (68.5 g, 95.8 g, 102.7 g and 143.8 g for 1.0, 1.4, 1.5, 2.1 M sucrose solutions, respectively) in H_2O. Add 2 mL of 0.1 M $NaHCO_3$ to each solution and make up to 200 mL with H_2O.

12. 1 mM $NaHCO_3$. Dilute 100 mM $NaHCO_3$ into H_2O. 400 mL/20 g of starting brain is required.

13. Plastic disposable pipettes. e.g., Liquipette, polyethylene transfer pipettes of 4 mL capacity, thin stem, 7 mL capacity, with scale, and 6 mL capacity 9″ long (Elkay, Shrewsbury, MA), or other plastic Pasteur pipette such as 3 mL (with scale).

2.2 Preparation of SPM and PSD Fraction (Long Procedure)

The long procedure requires solutions used in Subheading 2.1 and additional solutions listed below.

1. Sucrose solutions (0.85, 1.0 and 1. 2 M). Dissolve sucrose (58.2 g, 68.5 g, and 82.2 g for 0.85, 1.0, and 1. 2 M sucrose solutions, respectively) in H_2O, add 2 mL of 0.1 M $NaHCO_3$ and make up to 200 mL with H_2O.

2. 0.5 mM HEPES/KOH (pH 7.4). Dilute 0.5 M stock in H_2O. About 250–500 mL/20~25 g brain is required for SPM preparation.

3. 1 mM $NaHCO_3$. Dilute 100 mM $NaHCO_3$ into H_2O. About 150 mL/20–25 g of starting brain is required.

2.3 Preparation of S₃ and P₃ Fraction

No additional reagent or solution is necessary.

2.4 Preparation of Synaptic Membrane Raft from SPM

1. TNE buffer: 20 mM Tris–HCl (pH 7.4) containing 150 mM NaCl and 1 mM EDTA. Store at 4 °C.

2. 2× detergent stock solution in TNE buffer. Weigh and dissolve detergent in TNE buffer. Detergent concentration is weight/volume. Store at 4 °C. Examples of detergents include TX-100, OG and 3-([3-cholamidopropyl]dimethylammonio)-2-hydroxy-1-propanesulfonate (CHAPSO).

3. Sucrose solutions (80%, 30%, and 5%) in TNE buffer. Dissolve sucrose in TNE buffer (160 g, 60 g, and 10 g for 80%, 30%, and 5% sucrose solutions, respectively in 200 mL). Degas the sucrose solution by vacuum pump until no air bubbles form. Deaeration is required to prevent disturbance of the sucrose gradient caused by bubbles during sucrose density gradient (SDG) ultracentrifugation.

4. 1 M IAA. Weigh ~20 mg of IAA and add an appropriate volume of H_2O to make a 1 M solution. Mix at room temperature until no crystals remain. Let stand at room temperature before use. Prepare on the day of use.

5. Protease inhibitor cocktail (P8340, Sigma-Aldrich, St Louis, MO). Store the stock reagent at −30 °C. Thaw at room temperature on the day of use.

2.5 Preparation of PSD Lattice from SPM

No additional reagent or solution is necessary (*see* Subheading 2.4).

3 Methods

3.1 Preparation of P₁, P₂, Synaptosome, and PSD Fraction (Short Procedure) from Rat Forebrain

The method is based on those developed by Siekevitz's group [2–4]. Protocol for PSD purification (short procedure) using 20–25 g forebrain as starting material is described below. The maximum amount of forebrains is about 25 g due to the limitation of capacity of ultracentrifuge. All the processes are carried out at 4 °C. The procedure is outlined in Fig. 4.

1. Collect rat forebrains by decapitation and quick dissection (*see* **Note 3**). Place forebrains immediately after dissection in a beaker placed on ice. Weigh the pooled brains (weight of the container is better be measured before pooling tissues). Proceed for **step 2** or freeze and keep the forebrains at −80 °C until use.

2. Chop forebrains into small pieces (about less than 2 × 2 × 2 mm) with scissors. When using frozen brains, dip frozen brains into small amount of cooled solution A (~ a few mL) in a beaker, chop or scrape them by scissors. Add solution A to make

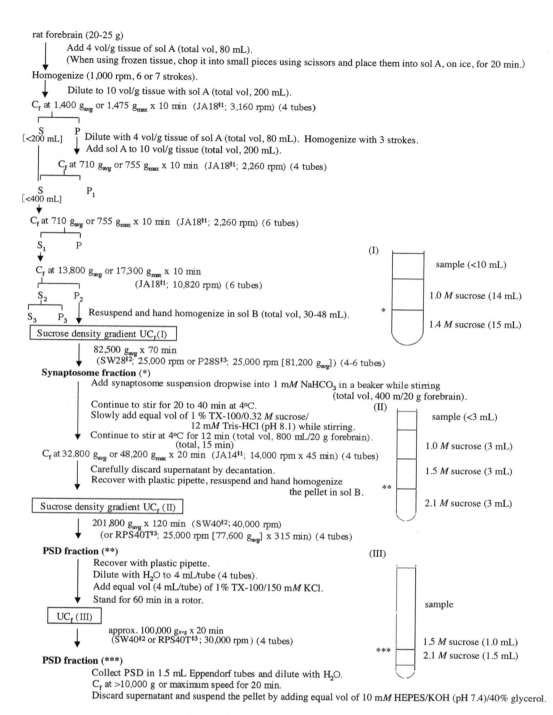

rat forebrain (20-25 g)

Add 4 vol/g tissue of sol A (total vol, 80 mL).
(When using frozen tissue, chop it into small pieces using scissors and place them into sol A, on ice, for 20 min.)

Homogenize (1,000 rpm, 6 or 7 strokes).

Dilute to 10 vol/g tissue with sol A (total vol, 200 mL).

C_f at 1,400 g_{avg} or 1,475 g_{max} x 10 min (JA18[‡1]; 3,160 rpm) (4 tubes)

S [<200 mL]   P   Dilute with 4 vol/g tissue of sol A (total vol, 80 mL). Homogenize with 3 strokes.
Add sol A to 10 vol/g tissue (total vol, 200 mL).

C_f at 710 g_{avg} or 755 g_{max} x 10 min (JA18[‡1]; 2,260 rpm) (4 tubes)

S [<400 mL] P_1

C_f at 710 g_{avg} or 755 g_{max} x 10 min (JA18[‡1]; 2,260 rpm) (6 tubes)

S_1 P

C_f at 13,800 g_{avg} or 17,300 g_{max} x 10 min
(JA18[‡1]; 10,820 rpm) (6 tubes)

S_2 P_2
S_3 P_3 Resuspend and hand homogenize in sol B (total vol, 30-48 mL).

Sucrose density gradient UC$_f$(I)

82,500 g_{avg} x 70 min
(SW28[‡2]; 25,000 rpm or P28S[‡3]; 25,000 rpm [81,200 g_{avg}]) (4-6 tubes)

Synaptosome fraction (*)

Add synaptosome suspension dropwise into 1 mM NaHCO$_3$ in a beaker while stirring
(total vol, 400 m/20 g forebrain).
Continue to stir for 20 to 40 min at 4°C.
Slowly add equal vol of 1 % TX-100/0.32 M sucrose/
12 mM Tris-HCl (pH 8.1) while stirring.
Continue to stir at 4°C for 12 min (total vol, 800 mL/20 g forebrain).
(total, 15 min)

C_f at 32,800 g_{avg} or 48,200 g_{max} x 20 min (JA14[‡1]; 14,000 rpm x 45 min) (4 tubes)

Carefully discard supernatant by decantation.
Recover with plastic pipette, resuspend and hand homogenize
the pellet in sol B.

Sucrose density gradient UC$_f$ (II)

201,800 g_{avg} x 120 min (SW40[‡2]; 40,000 rpm)
(or RPS40T[‡3]; 25,000 rpm [77,600 g_{avg}] x 315 min) (4 tubes)

PSD fraction (**)

Recover with plastic pipette.
Dilute with H$_2$O to 4 mL/tube (4 tubes).
Add equal vol (4 mL/tube) of 1% TX-100/150 mM KCl.
Stand for 60 min in a rotor.

UC$_f$ (III)

approx. 100,000 g_{avg} x 20 min
(SW40[‡2] or RPS40T[‡3]; 30,000 rpm) (4 tubes)

PSD fraction (***)

Collect PSD in 1.5 mL Eppendorf tubes and dilute with H$_2$O.
C_f at >10,000 g or maximum speed for 20 min.
Discard supernatant and suspend the pellet by adding equal vol of 10 mM HEPES/KOH (pH 7.4)/40% glycerol.

(I)
sample (<10 mL)
1.0 M sucrose (14 mL)
*
1.4 M sucrose (15 mL)

(II)
sample (<3 mL)
1.0 M sucrose (3 mL)
1.5 M sucrose (3 mL)
**
2.1 M sucrose (3 mL)

(III)
sample
1.5 M sucrose (1.0 mL)
***
2.1 M sucrose (1.5 mL)

Fig. 4 Purification of synaptosome and PSD by subcellular fractionation using sucrose density gradient centrifugation (PSD purification by short procedure). Examples of centrifugation conditions (rotors and speed) are indicated. Rotors marked with [‡1], [‡2] and [‡3] are those for centrifuges of Avanti J-25 (Beckman), L5–50 (Beckman) and Himac CP60E (Hitachi), respectively. Steps using ultracentrifuge are numbered with roman characters and surrounded with square. Volumes and number of centrifuge tubes used are those for purifying PSD from 20–25 g forebrains of rats. *, ** and *** Positions where synaptosome and PSD before and after TX-100/KCl treatment, respectively, are collected. C_f centrifugation, UC_f ultracentrifugation, *av*, average, *sol* solution, *vol* volume

80 mL suspension. Keep the suspension on ice for at least 20 min when using frozen brains (*see* **Note 4**).

3. Homogenize the suspension at 1000 rpm with 6 or 7 up-and-down motions with a motor-operated Teflon/glass homogenizer using a loose-fitting pestle (*see* **Note 5**) while cooling the container in ice water. Recover suspension into a new beaker and dilute to 200 mL with solution A. (Start preparing sucrose layers necessary at **step 9** during centrifugations at **steps 3** (or **4**)–**8**.)

4. Centrifuge at 1400 × g_{av} or 1475 × g_{max} for 10 min (JA18; 3160 rpm, 4 tubes). Save supernatant in a beaker placed on ice or at 4 °C.

5. Dilute the pellet with solution A and make 80 mL suspension. Homogenize with 3 strokes as in **step 3**. Dilute with solution A to 200 mL.

6. Centrifuge at 710 × g_{av} or 755 × g_{max} for 10 min (JA18; 2260 rpm, 4 tubes). Collect supernatant. Pellet is P_1.

7. Combine supernatants obtained in **steps 4** and **6** and centrifuge at 710 × g_{av} or 755 × g_{max} for 10 min (JA18; 2260 rpm, 6 tubes) (*see* **Note 6**).

8. Collect supernatant (S_1) and centrifuge at 13,800 × g_{av} or 17,300 × g_{max} for 10 min (JA18; 10,820 rpm, 6 tubes). Supernatant and pellet obtained in this step are S_2 and P_2, respectively.

9. Resuspend the P_2 and gently hand homogenize with a Dounce homogenizer or Teflon-glass homogenizer in solution B (~48 mL). Layer the suspension on gradients composed of 1.0 and 1.4 M sucrose, and centrifuge at 82,500 × g_{av} for 70 min (SW 28, 25,000 rpm, 4–6 tubes) (*see* **Note 7**).

10. Collect the bands in the interface between 1.0 and 1.4 M sucrose layer (synaptosome fraction) (*see* **Note 8**) into a small beaker with a plastic pipette of 4 mL capacity with thin stem (*see* **Note 9**). Measure the volume of the synaptosome suspension, if necessary. Protein concentration of synaptosome just after recovery from the interface band is approx. 5 mg protein/mL. Save aliquot of synaptosome suspension after dilution to make about 2.5 mg/mL (just an example), if necessary.

11. Pour 1 mM $NaHCO_3$ into a large beaker to make the final volume after mixing of the synaptosome suspension 400 mL/20 g starting forebrains (*see* **Note 10**). Place a stirrer bar into the beaker. Add synaptosome suspension dropwise into 1 mM $NaHCO_3$ in a beaker while stirring. Continue to stir for about 20–40 min at 4 °C (*see* **Note 11**).

12. Add slowly 400 mL/20 g of starting forebrains of 1% TX-100/0.32 M sucrose/12 mM Tris–HCl (pH 8.1) (final 0.5% TX-100, 0.16 M sucrose, 6 mM Tris–HCl) with constant stirring. Take 1 min to add the TX-100 solution. Continue to stir at 4 °C. Total time of treatment with TX-100 (from starting addition of TX-100 to starting next centrifugation) should be 15 min. Therefore, transfer the solution to the transparent centrifuge tubes (*see* **Note 12**) at about 12 min after starting addition of TX-100 (*see* **Note 13**).

13. Centrifuge at $32,800 \times g_{av}$ or $48,200 \times g_{max}$ for 20 min (JA14; 14,000 rpm × 45 min, four 250 mL tubes). (Prepare sucrose layers required at **step 14** by using a plastic Pasteur pipette with scale.) Discard upper large portion of supernatant by slow decantation. Discard supernatant using a plastic pipette of 6 mL with 9″-long so that about 2 mL supernatant remains in the tube. Be very careful not to disturb the pellet. Recover pellet with plastic pipette by peeling and aspirating the pellet as a mass. Collect the pellet as small a volume as possible. Resuspend the pellet in solution B. Gently hand homogenize the pellet with a Dounce homogenizer or loose Teflon-glass homogenizer.

14. Layer the solution on gradients composed of 1.0, 1.5, and 2.1 M sucrose, and centrifuge at $201,800 \times g_{av}$ for 120 min (SW40; 40,000 rpm or RPS40T; 25,000 rpm 315 min, 4 tubes) (*see* **Notes 14** and **15**). (In the latter case, next step begins next morning.)

15. Recover PSD fraction (∗∗) with a plastic pipette (4 mL with thin stem) into 15 mL plastic tube. Dilute with cold H_2O to 4 mL/1 tube and mix homogeneously. Add equal volume [4 mL/tube] of 1% TX-100/150 mM KCl (final 0.5% TX-100, 75 mM KCl) and mix homogeneously. Stand for 60 min (*see* **Note 16**).

16. Layer the solution on gradients composed of 1.5 and 2.1 M sucrose, and centrifuge at approx. $100,000 \times g_{av}$ for 20 min (SW40 or RPS40T; 30,000 rpm, 2 tubes) (*see* **Note 14**).

17. Retrieve PSD fraction (∗∗∗) with a plastic pipette (4 mL with thin stem) into 1.5 mL Eppendorf microfuge tubes. Dilute with more than an equal volume of cold H_2O (*see* **Note 17**). Centrifuge at $>10,000 \times g$ for 20 min. (Swing rotor is favorable).

18. Discard supernatant and weigh the PSD material. Add equal amount of 10 mM HEPES/KOH (pH 7.4)/40% glycerol and mix homogeneously (*see* **Note 18**). Divide into small aliquots and keep them in plastic tubes at −80 °C until use.

3.2 Preparation of SPM and PSD Fraction (Long Procedure)

Protocol (long procedure) for PSD purification using 20–25 g forebrain as starting material is described below. All the processes are carried out at 4 °C. The procedure is outlined in Fig. 5. **Steps 6–12** are the same as **steps 12–18** of short procedure except for volumes of the samples and the number of centrifuge tubes used.

1. Prepare synaptosome fraction following the protocol described in Subheading 3.1.

2. Pour 0.5 mM HEPES/KOH (pH 7.4) into a large beaker to make the final volume after mixing the synaptosome suspension 400 mL. Place a stirrer bar into the beaker. Add synaptosome suspension (~50 mL) dropwise into the HEPES/KOH buffer in a beaker while stirring. Continue to stir for about 45 min at 4 °C.

3. Centrifuge at 32,800 × g_{av} or 48,200 × g_{max} for 20 min (JA14; 14,000 rpm × 45 min). Collect pellet and resuspend in solution B as stated in Subheading 3.1, **step 13**.

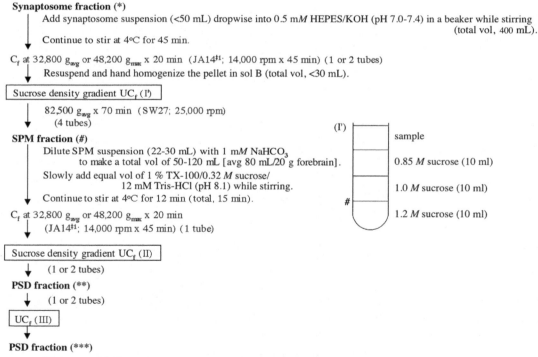

Synaptosome fraction (*)

　Add synaptosome suspension (<50 mL) dropwise into 0.5 mM HEPES/KOH (pH 7.0-7.4) in a beaker while stirring (total vol, 400 mL).
　Continue to stir at 4°C for 45 min.

C_f at 32,800 g_{avg} or 48,200 g_{max} x 20 min (JA14$^{\#1}$; 14,000 rpm x 45 min) (1 or 2 tubes)
　Resuspend and hand homogenize the pellet in sol B (total vol, <30 mL).

Sucrose density gradient UC_f (I')

　82,500 g_{avg} x 70 min (SW27; 25,000 rpm)
　(4 tubes)

(I')
sample
0.85 M sucrose (10 ml)
1.0 M sucrose (10 ml)
1.2 M sucrose (10 ml)

SPM fraction (#)

　Dilute SPM suspension (22-30 mL) with 1 mM NaHCO$_3$ to make a total vol of 50-120 mL [avg 80 mL/20 g forebrain].
　Slowly add equal vol of 1 % TX-100/0.32 M sucrose/ 12 mM Tris-HCl (pH 8.1) while stirring.
　Continue to stir at 4°C for 12 min (total, 15 min).

C_f at 32,800 g_{avg} or 48,200 g_{max} x 20 min
　(JA14$^{\#1}$; 14,000 rpm x 45 min) (1 tube)

Sucrose density gradient UC_f (II)

　(1 or 2 tubes)

PSD fraction (**)

　(1 or 2 tubes)

UC_f (III)

PSD fraction (***)

　Wash once with H$_2$O and suspend PSD in 5 mM HEPES-KOH/20% glycerol.

Fig. 5 Purification of SPM and PSD by subcellular fractionation using sucrose density gradient centrifugation (PSD purification by long procedure). Protocol to prepare synaptosome and steps after sucrose gradient centrifugation (II) are essentially the same as those shown in Fig. 4. Comments and abbreviations are the same as in Fig. 4

4. Layer the suspension on gradients composed of 0.85, 1.0 and 1.2 M sucrose, and centrifuge at $82,500 \times g_{av}$ for 70 min (SW 28, 25,000 rpm). Use 4 tubes.

5. Collect SPM in the 1.0–1.2 M sucrose interface (#). Volume of this suspension is usually 20–30 mL (*approx. 3–4 mg protein/mL).

6. For subsequent purification of PSD, dilute SPM suspension with 1 mM NaHCO$_3$ (final vol, 50–120 mL [average 80 mL/20 g forebrain]).

7. Treat the SPM suspension by adding equal volume of TX-100 as stated in Subheading 3.1, **step 12**.

8. Centrifuge at $32,800 \times g_{av}$ or $48,200 \times g_{max}$ for 20 min (JA14; 14,000 rpm × 45 min, 1 tube). Collect pellet and resuspend in solution B as stated in Subheading 3.1, **step 13**.

9. Layer the solution on the top of the sucrose gradient and centrifuge at $201,800 \times g_{av} \times 120$ min (SW40, 40,000 rpm or RPS40T; 25,000 rpm 315 min, 1 tube).

10. Recover PSD fraction (**) as stated in Subheading 3.1, **step 15**.

11. Centrifuge at approx. $100,000 \times g_{av}$ for 20 min (SW40 or RPS40T; 30,000 rpm, 1 tube) as stated in Subheading 3.1, **step 16**.

12. Retrieve PSD fraction (***), process and save as stated in Subheading 3.1, **steps 17** and **18**.

3.3 Preparation of S_3 and P_3 Fraction

Centrifuge S_2 material at $100,000 \times g$ for 1 h. Supernatant and pellet obtained are S_3 and P_3 fractions, respectively.

3.4 Preparation of Synaptic Membrane Raft from SPM

The protocol provided in this chapter is useful for the purification of the synaptic membrane raft, simultaneous purification of the PSD, and investigation of relationship between the synaptic membrane raft and the PSD. It is desirable to use SPM prepared in the presence of 2 mM IAA. Both freshly prepared SPM and SPM stored at −30 °C in the presence of 50% glycerol can be used. All processes are carried out at 4 °C. The standard protocol, outlined in Fig. 6, uses 500 µg of SPM protein. Typically, 20 and 1.5 µL of each fraction are used for the SDS-PAGE and GM1 immuno-dot blot, respectively. Therefore, more than 40 repetitions of SDS-PAGE (stained with silver or SYPRO Ruby) and western blotting can be carried out.

Researchers can modify the protocol by using different types of detergents and various concentrations of detergent (typically ranging from 0.05–5%) (*see* **Note 19**). The protein profile of SDG and the protein components in the membrane raft and PSD fractions differ depending on the detergent used and the ratio of detergent

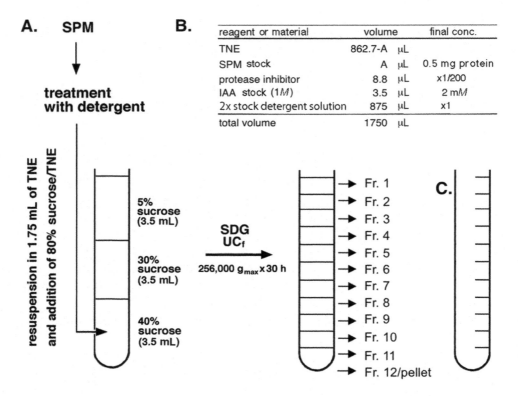

A. SPM

treatment
with detergent

resuspension in 1.75 mL of TNE
and addition of 80% sucrose/TNE

5%
sucrose
(3.5 mL)

30%
sucrose
(3.5 mL)

40%
sucrose
(3.5 mL)

**SDG
UC$_f$**

256,000 g$_{max}$ × 30 h

→ Fr. 1
→ Fr. 2
→ Fr. 3
→ Fr. 4
→ Fr. 5
→ Fr. 6
→ Fr. 7
→ Fr. 8
→ Fr. 9
→ Fr. 10
→ Fr. 11
→ Fr. 12/pellet

C.

B.

reagent or material	volume	final conc.
TNE	862.7-A µL	
SPM stock	A µL	0.5 mg protein
protease inhibitor	8.8 µL	×1/200
IAA stock (1M)	3.5 µL	2 mM
2x stock detergent solution	875 µL	×1
total volume	1750 µL	

Fig. 6 Purification protocol of synaptic membrane rafts by sucrose density gradient centrifugation. (**a**) Protocol using the SPM as the starting material. (**b**) Example of detergent treatment. The total SPM protein amount during detergent treatment is 0.5 mg. Change the types and concentration of detergent as required. Detergent treatment is carried out at 4 °C for 30 min with gentle rotational mixing. (**c**) Centrifuge tube marked with the volumes of each fraction (955 µL for each fraction). Mark after every addition of 955 µL H$_2$O

to protein. Typical examples of SDG protein distribution are shown in Fig. 7. The potency of the holding raft-PSD complex differs with detergent type and their concentrations. For example, low concentration TX-100 (e.g., 0.15%) maintains the membrane raft–PSD complex [38, 41]. OG dissociates the synaptic membrane raft and the PSD [42]. OG and CHAPSO tend to maintain the membrane raft integrity better than TX-100 [43]. Fraction 12/pellet contains the PSD [41]. Fractions 8–11 are mixtures of soluble and cytoskeletal proteins. The membrane rafts are typically distributed in fractions 4–6 after treatment with 0.15% TX-100 (Fig. 7) [38, 41, 43]. Identify the membrane raft-containing fractions by GM1 immuno–dot blot assay.

1. Mix SPM (0.5 mg protein) (*see* **Note 20**), TNE buffer, protease inhibitor cocktail, and IAA in a 15 mL screw-cap plastic tube as indicated in Fig. 6b. Mix gently but well.

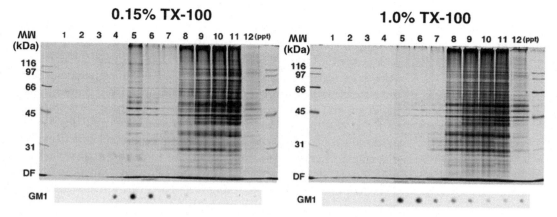

Fig. 7 Electrophoretic profile of detergent-treated SPM after SDG ultracentrifugation. Representative examples are shown. Proteins were stained with silver. The lower images are the GM1 distribution on the sucrose gradient revealed by dot blot analysis using horseradish peroxidase-conjugated cholera toxin B subunit, which specifically binds to GM1 ganglioside, a membrane raft marker. Membrane rafts are distributed on fractions 4–6 in the 0.15% TX-100 treatment. Fraction 12/pellet contains the PSD. It should be noted that the raft fractions in the 0.15% TX-100 treatment contain the synaptic membrane raft-PSD complexes [38, 41]. Fractions 8–11 are a mixture of soluble and cytoskeletal proteins. The protein distribution profiles on the SDG depend on the detergent-to-protein ratio. In this protocol, the detergent-to-protein ratios (w/w) are 5.25 and 35, respectively, for the 0.15% and 1.0% TX-100 treatments. Fraction numbers are shown at the tops of the gels, and the molecular weights (MW, in kDa) are on the left. DF and ppt refer to the dye front and pellet, respectively. The images were extracted from a previous publication [41]

2. Add 875 μL of 2× stock detergent solution to the tube. Close the screw cap of the tube, mix well, and continue gentle mixing using a rotator for 30 min.

3. Prepare 1.75 mL 80% sucrose/TNE (2.3 g) in an ultracentrifuge tube while detergent treatment is underway.

4. Transfer the detergent-treated sample to the ultracentrifuge tube containing 1.75 mL 80% sucrose/TNE and mix well.

5. Gently overlay the TNE buffers containing 30% sucrose and 5% sucrose sequentially without disturbing the interfaces of the sucrose solutions.

6. Centrifuge at 256,000 × g_{max} for 30 h.

7. Collect the 11 fractions (955 μL for each fraction) from the top using a 1 mL pipette. Follow the solution levels using the pre-marked lines on a second ultracentrifuge tube (Fig. 6c) (*see* **Note 21**). Collect fraction 12/pellet by repeated flushing with the recovery solution by pipette (*see* **Note 22**). Mix each fraction, except for fraction 12, gently by brief vortex to make the solutions homogeneous (*see* **Note 23**).

3.5 Preparation of PSD Lattice from SPM

Purification and characterization of PSD lattice, presumably a backbone structure for excitatory PSD in the mammalian central nervous system, was reported in 2018 [47]. However, the preparation, and therefore, the protocol had disadvantage, in particular, for the identification of protein composition, because the preparation contained not a few SDS-insoluble proteins. After the initial paper on the PSD lattice [47], the initial protocol was improved and the insolubility problem was solved. The new PSD lattice preparation is considered to be more physiological than the previous one and enabled identification of the component proteins by SDS-PAGE and western blotting (Suzuki et al., in preparation). This chapter describes the improved protocol. A protocol for purification of the PSD lattice in the forebrain SPM shown in Fig. 8 uses 3 mg of forebrain SPM protein as a starting material. Use SPM prepared in the presence of 2 mM IAA. First half of the protocol is the same as that for synaptic membrane raft purification. All processes are carried out at 4 °C. Detergents other than OG were not tested.

1. Incubate the SPM protein (3 mg) (see **Note 20**) in TNE buffer containing protease inhibitor cocktail, 2 mM IAA, and 1%OG in a 50 mL screw-cap plastic tube to make a total volume of 10.5 mL (1.75 mL ×6) (*see* **Note 24**). Close the screw cap of the tube, mix well, and continue gentle mixing using a rotator for 30 min.

2. Mix to the detergent-treated solution with equal volume (10.5 mL) of TNE buffers containing 80% sucrose and devide the mixture into 6 ultracentrifugation tubes. Overlay with TNE buffers containing 30% sucrose and then 5% sucrose (Each tube contains 3 layers of 3.5 mL sucrose solution), and centrifuge at $256,000 \times g_{max}$ for 30 h at 4 °C.

3. Slowly aspirate the solution localizing in the fractions 1–10 from the top, using a plastic pipette by following the premarked volume lines on a second tube (Fig. 6c) (*see* **Note 25**).

4. Collect upper and lower portions of fraction 11 (1%OG-11U and1%OG-11B, respectively) (typically 825 μLD and 130 μL, respectively) separately (*see* **Notes 26** and **27**). Resuspend the pellet (1%OG-12) in 955 μL of 5 mM HEPES/KOH (pH 7.4) containing 50%glycerol.

5. Dilute 1%OG-11U and 1%OG-11B by 4 fold with 5 mM HEPES/KOH (pH 7.4), centrifuge at $100,000 \times g_{max}$ for 30 min at 4 °C. Resuspend the pellets in 1 mL of 5 mM HEPES/KOH (pH 7.4), ultracentrifuge again, and resuspend the final pellets (1%OG-11U-IS and 1%OG-11B-IS, respectively) in 100 μL of 5 mM HEPES/KOH (pH 7.4) containing 50%glycerol (*see* **Notes 28** and **29**).

6. The preparations were stored unfrozen at −30 °C.

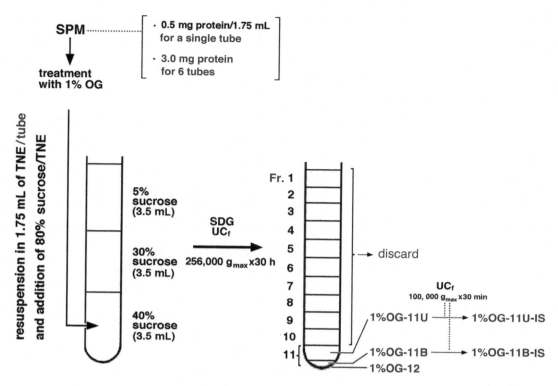

Fig. 8 The purification protocol of the PSD lattice by sucrose density gradient centrifugation. The protocol shown is an improved version of the previous one [47]. (Suzuki et al., in preparation). First half of the protocol is the same as that for synaptic membrane raft purification until the process of SDG ultracentrifugation shown in Fig. 6. We refer 1%OG-11U-IS as PSD lattice preparation. Protein components in these two preparations are nearly the same as far as SDS-PAGE profiles are compared (Suzuki et al., in preparation). U, B, IS, and UC_{ff} refer to upper, bottom, insoluble, and ultracentrifugation, respectively

4 Notes

1. Subtle changes in ionic strength and metal concentration may affect sedimentation of subcellular organelles and protein complexes. It is not necessary to use ultrapure water, such as nanopure or miliQ water, for this subfractionation, and the usage of ultrapure water may sometimes result in low yield of synaptosome and PSDs. Some unidentified factor(s) affects the sedimentation and/or are necessary for stabilization of PSD protein complex.

2. Commercial TX-100 has been found to contain impurity with oxidizing activity [48].

3. If brains are homogenized or rapidly frozen in liquid nitrogen within 30 s to 1 min after decapitation, content of CaMKII, both α and β, are very low in the PSD fraction [22]. Neurofilament content is increased in such PSD preparation.

4. Defrozen and chopped brains should be kept in cooled solution A for at least 20 min to depolymerize actin cytoskeleton. Inadequate depolymerization causes unfavorable sedimentation.

5. Literature [1, 2] recommends loose homogenizer (e.g., Teflon-glass homogenizer with a clearance of 0.25 mm or Dounce homogenizer with a loose-fitting pestle) to preserve morphological integrity of PSD. However, 0.25 mm clearance homogenizer does not appear to be a must.

6. It is very difficult to separate clearly the supernatant and pellet from total brain homogenate by centrifugation at $755 \times g_{max}$. Therefore, the first centrifugation was carried out at $1475 \times g_{max}$. Supernatant obtained in the first centrifugation and the second centrifugation at $755 \times g_{max}$ are combined, centrifuged again at $755 \times g_{max}$, and thus S_1 fraction was obtained. Removing $755 \times g_{max}$ pellet is important to minimize contamination of nuclear materials to synaptic fractions [49]. Methods omitting this step (e.g., one-step purification of synaptosome) cannot avoid large amounts of contamination of nuclear proteins.

7. The first sucrose gradient was originally composed of 0.85, 1.0, and 1.2 M sucrose [2, 4], but was replaced by those composed of 1.0 and 1.4 M sucrose with equivalent result [3].

8. Use fresh unfrozen brain as starting tissues for functional analysis of synaptosome. It is required to incubate synaptosome suspension in normotonic buffer to bring the terminals to a physiological steady state [50]. Synaptosomes recovered from the sucrose gradient and are not incubated in normotonic buffer are shrunken due to high osmotic pressure.

9. Using disposable plastic pipettes to collect synaptosome, SPM, and PSD enriched bands after sucrose gradient centrifugation is convenient. See also Subheading 2.1, **item 13**. Be careful not to warm the plastic pipette (it means protein sample) by holding it with warm hand or fingers with wide contact areas for long time.

10. Fixed volume (400 mL for 20 g starting tissue) of the synaptosome suspension just before the TX-100 treatment is based on the protein concentration [2] estimated by the A_{260} and A_{280} using nomogram (distributed by California Corporation for Biochemical Research, LA) based on the equation by Warburg and Christian [51]. Dilute synaptosome solution by 40 times for measurement of A_{260} and A_{280}. (This also applies to SPM solution). The protein concentration estimated by Warburg-Christian method is about fourfolds of the value obtained by Lowry method using BSA as standard. (The values were

$4.3 \pm 1.2 \, [n = 7]$ and $3.3 \pm 1.2 \, [n = 12]$ folds for synaptosome and SPM fractions, respectively.) Therefore, protein concentration of the synaptosome suspension in 400 mL/20 g original forebrain is approximately 1 mg protein/mL (not 4 mg protein/mL as written in the original paper). Volume should be changed when starting from other parts of the brain, such as cerebellum.

11. This process is required before treatment with TX-100 and important to obtain good yield of PSD proteins, although the reason is unknown. Omitting this process may bring low yields of PSD.

12. Use transparent centrifuge tube to see the pellet clearly with the naked eye. The pellet obtained is very soft and easy to disturb. It is required to collect the pellet in a small volume to load on the top layer of the next sucrose density gradient.

13. The duration of TX-100 treatment affects the recovery of PSD.

14. PSD is extremely sticky to glass and cellulose nitrate tubes [2]. Use polyallomer centrifuge tubes [2] to prevent adhere PSD to the tubes.

15. Keep temperature to be around 4 °C during ultracentrifugation. Raise of temperature loses some enzyme activity.

16. Inadequate treatment at this step leaves membrane materials to the final PSD preparation.

17. Repeat wash once or twice if complete removal of TX-100 is required.

18. Glycerol should be added to prevent artificial aggregation of the PSD proteins during storage at −80 °C. Again, PSD material is extremely sticky to glass and cellulose nitrate, and tend to aggregate very easily, in particular, after freezing and defreezing.

19. Consider the critical micelle concentration (CMC) of the detergents. In particular, the CMC of OG is high (20–25 mM/0.585–0.7%).

20. Use SPM prepared in the presence of 2 mM IAA, because the subsequent protocol is carried out in the presence of IAA.

21. Collection using a 1 mL pipette is unreliable because the sucrose density differs from fraction to fraction (the gradient ranges from 5% to 40% from the top of the tube to the bottom), resulting in the recovery of different volumes by pipette. Take advantage of the pre-marked levels on a second ultracentrifuge centrifuge tube for this step.

22. Vigorous vortexing of the PSD-containing solution may cause denaturation of the PSD proteins and increase insolubility of the PSD protein complex. The recovery solution should contain 40–50% glycerol to prevent irreversible aggregation of PSDs (*see* also **Note 18**).

23. The solution recovered by pipette from the sucrose gradient is not homogeneous even in small fractions. Therefore, mixing by brief vortex is required. Avoid vortexing fraction 12/pellet, which contains the PSD.

24. This amount is sixfold of the standard-scale purification protocol for synaptic membrane raft shown in Fig. 6.

25. Positions of 11 fractions (955 μL each) were marked on a second centrifuge tube and numbered from the top, as shown in Fig. 6.

26. Pre-mark the surface level of 130 μL on a second centrifuge tube after the addition of 130 μL H_2O. Be very careful not to contaminate the PSDs in the 1%OG-IS-11B/PSD lattice fraction by touching the tip of pipette with the PSD-containing pellet.

27. The volume of bottom portion (therefore, that of upper portion, too) critically affects on the protein recovery and sparseness/density of the PSD lattice structure in the upper portion. It is better not to collect bottom portion more than 200 μL. Otherwise, protein recovery may be too low for subsequent biochemical analyses. More densely packed PSD lattice is recovered in the more bottom portion.

28. The pellet materials were not visible with the naked eye during the last purification process.

29. Collect 50 μL of suspension twice and make final volume 100 μL. The final solutions were not hand-homogenized nor vortexed to avoid any loss and protein denaturation.

Acknowledgments

The author learned the method of PSD purification in the Philip Siekevitz laboratory, Rockefeller University, New York. The author heartily thanks Dr. Philip Siekevitz and Marie LeDoux for their instruction.

References

1. Somerville RA, Merz PA, Carp RI (1984) The effects of detergents on the composition of postsynaptic densities. J Neurochem 43(1):184–191

2. Cohen RS et al (1977) The structure of post-synaptic densities isolated from dog cerebral cortex. I. Overall morphology and protein composition. J Cell Biol 74(1):181–203

3. Wu K, Carlin R, Siekevitz P (1986) Binding of L-[3H]glutamate to fresh or frozen synaptic membrane and postsynaptic density fractions isolated from cerebral cortex and cerebellum of fresh or frozen canine brain. J Neurochem 46(3):831–841

4. Carlin RK et al (1980) Isolation and characterization of postsynaptic densities from various brain regions: enrichment of different types of postsynaptic densities. J Cell Biol 86(3):831–845

5. Kim TW, Wu K, Black IB (1995) Deficiency of brain synaptic dystrophin in human Duchenne muscular dystrophy. Ann Neurol 38(3):446–449

6. Hahn CG et al (2009) The post-synaptic density of human postmortem brain tissues: an experimental study paradigm for neuropsychiatric illnesses. PLoS One 4(4):e5251

7. Suzuki T et al (1993) Characterization of protein kinase C activities in postsynaptic density fractions prepared from cerebral cortex, hippocampus, and cerebellum. Brain Res 619(1–2):69–75

8. Kim TW et al (1992) Detection of dystrophin in the postsynaptic density of rat brain and deficiency in a mouse model of Duchenne muscular dystrophy. Proc Natl Acad Sci U S A 89(23):11642–11644

9. Wu K, Black IB (1987) Regulation of molecular components of the synapse in the developing and adult rat superior cervical ganglion. Proc Natl Acad Sci U S A 84(23):8687–8691

10. Wu K, Siekevitz P (1988) Neurochemical characteristics of a postsynaptic density fraction isolated from adult canine hippocampus. Brain Res 457(1):98–112

11. Suzuki T et al (1997) Excitable membranes and synaptic transmission: postsynaptic mechanisms. Localization of alpha-internexin in the postsynaptic density of the rat brain. Brain Res 765(1):74–80

12. Matus A et al (1980) Brain postsynaptic densities: the relationship to glial and neuronal filaments. J Cell Biol 87(2 Pt 1):346–359

13. Suzuki T et al (2007) Characterization of mRNA species that are associated with postsynaptic density fraction by gene chip microarray analysis. Neurosci Res 57(1):61–85

14. Cotman CW, Taylor D (1972) Isolation and structural studies on synaptic complexes from rat brain. J Cell Biol 55(3):696–711

15. Nieto-Sampedro M, Bussineau CM, Cotman CW (1981) Optimal concentration of iodonitrotetrazolium for the isolation of junctional fractions from rat brain. Neurochem Res 6(3):307–320

16. Cotman CW et al (1974) Isolation of postsynaptic densities from rat brain. J Cell Biol 63(2 Pt 1):441–455

17. Kelly PT, Montgomery PR (1982) Subcellular localization of the 52,000 molecular weight major postsynaptic density protein. Brain Res 233(2):265–286

18. Kelly PT, Cotman CW (1976) Intermolecular disulfide bonds at central nervous system synaptic junctions. Biochem Biophys Res Commun 73(4):858–864

19. Kelly PT, Cotman CW (1981) Developmental changes in morphology and molecular composition of isolated synaptic junctional structures. Brain Res 206(2):251–257

20. Lai SL et al (1999) Interprotein disulfide bonds formed during isolation process tighten the structure of the postsynaptic density. J Neurochem 73(5):2130–2138

21. Sui CW, Chow WY, Chang YC (2000) Effects of disulfide bonds formed during isolation process on the structure of the postsynaptic density. Brain Res 873(2):268–273

22. Suzuki T et al (1994) Rapid translocation of cytosolic Ca2+/calmodulin-dependent protein kinase II into postsynaptic density after decapitation. J Neurochem 63(4):1529–1537

23. Carlin RK, Grab DJ, Siekevitz P (1982) Postmortem accumulation of tubulin in postsynaptic density preparations. J Neurochem 38(1):94–100

24. Cheng HH et al (2009) Cold-induced exodus of postsynaptic proteins from dendritic spines. J Neurosci Res 87(2):460–469

25. Li X et al (2007) Two pools of Triton X-100-insoluble GABA(A) receptors are present in the brain, one associated to lipid rafts and another one to the post-synaptic GABAergic complex. J Neurochem 102(4):1329–1345

26. Ratner N, Mahler H (1983) Isolation of postsynaptic densities retaining their membrane attachment. Neuroscience 9(3):631–644

27. Cho KO, Hunt CA, Kennedy MB (1992) The rat brain postsynaptic density fraction contains a homolog of the Drosophila discs-large tumor suppressor protein. Neuron 9(5):929–942

28. Walikonis RS et al (2000) Identification of proteins in the postsynaptic density fraction by mass spectrometry. J Neurosci 20(11):4069–4080

29. Murphy JA, Jensen ON, Walikonis RS (2006) BRAG1, a Sec7 domain-containing protein, is a component of the postsynaptic density of excitatory synapses. Brain Res 1120(1):35–45

30. Suzuki T (2002) Lipid rafts at postsynaptic sites: distribution, function and linkage to postsynaptic density. Neurosci Res 44(1):1–9

31. Blomberg F, Cohen RS, Siekevitz P (1977) The structure of postsynaptic densities isolated from dog cerebral cortex. II. Characterization and arrangement of some of the major proteins within the structure. J Cell Biol 74(1):204–225

32. Matus AI, Taff-Jones DH (1978) Morphology and molecular composition of isolated postsynaptic junctional structures. Proc R Soc Lond B Biol Sci 203(1151):135–151

33. Gurd JW, Gordon-Weeks P, Evans WH (1982) Biochemical and morphological comparison of postsynaptic densities prepared from rat, hamster, and monkey brains by phase partitioning. J Neurochem 39(4):1117–1124

34. Matus A (1981) The postsynaptic density. Trends Neurosci 4:51–53

35. Garner AE, Smith DA, Hooper NM (2008) Visualization of detergent solubilization of membranes: implications for the isolation of rafts. Biophys J 94(4):1326–1340

36. Shogomori H, Brown DA (2003) Use of detergents to study membrane rafts: the good, the bad, and the ugly. Biol Chem 384(9):1259–1263

37. Phillips GR et al (2001) The presynaptic particle web: ultrastructure, composition, dissolution, and reconstitution. Neuron 32(1):63–77

38. Suzuki T, Yao W-D (2014) Molecular and structural bases for postsynaptic signal processing: interaction between postsynaptic density and postsynaptic membrane rafts. J Neuro-Oncol 2:1–14

39. Pike LJ (2006) Rafts defined: a report on the Keystone Symposium on lipid rafts and cell function. J Lipid Res 47(7):1597–1598

40. Simons K, Ikonen E (1997) Functional rafts in cell membranes. Nature 387(6633):569–572

41. Suzuki T et al (2011) Association of membrane rafts and postsynaptic density: proteomics, biochemical, and ultrastructural analyses. J Neurochem 119(1):64–77

42. Liu Q, Yao W-D, Suzuki T (2013) Specific interaction of postsynaptic densities with membrane rafts isolated from synaptic plasma membranes. J Neurogenet 27(1–2):43–58

43. Zhao L, Sakagami H, Suzuki T (2014) Detergent-dependent separation of postsynaptic density, membrane rafts and other subsynaptic structures from the synaptic plasma membrane of rat forebrain. J Neurochem 131(2):147–162

44. Dosemeci A et al (2016) The postsynaptic density: there is more than meets the eye. Front Synaptic Neurosci 8:23

45. Matus AI, Walters BB (1975) Ultrastructure of the synaptic junctional lattice isolated from mammalian brain. J Neurocytol 4(3):369–375

46. Suzuki T et al (2017) Protein components of postsynaptic density lattice, a backbone structure for type I excitatory synapses. J Neurochem 144(4):390–407. in press

47. Suzuki T et al (2018) Protein components of post-synaptic density lattice, a backbone structure for type I excitatory synapses. J Neurochem 144(4):390–407

48. Chang HW, Bock E (1980) Pitfalls in the use of commercial nonionic detergents for the solubilization of integral membrane proteins: sulfhydryl oxidizing contaminants and their elimination. Anal Biochem 104(1):112–117

49. Adam RM et al (2008) Rapid preparation of nuclei-depleted detergent-resistant membrane fractions suitable for proteomics analysis. BMC Cell Biol 9:30

50. Fried RC, Blaustein MP (1978) Retrieval and recycling of synaptic vesicle membrane in pinched-off nerve terminals (synaptosomes). J Cell Biol 78(3):685–700

51. Warburg O, Christian W (1942) Isolierung und Kristallisation des Garungsferment. Biochem Z 310:384–421

Chapter 4

Laser Microdissection for Spatial Proteomics of Postmortem Brain Tissue

David C. Hondius, Jeroen J. M. Hoozemans, Annemieke J. M. Rozemuller, and August B. Smit

Abstract

Quantitative data of the proteome is highly valuable for providing unbiased information on protein expression changes related to disease or identifying related biomarkers. In brain diseases the affected area can be small and pathogenic events can be related to a specific cell type in an otherwise heterogeneous tissue type. An emerging approach dedicated to analysing this type of samples is laser microdissection (LMD) combined with LC-MS/MS into a single workflow. In this chapter, we describe two LMD methods for isolating tissue at the level of a small area and individual cells suitable for subsequent LC-MS/MS analysis.

Key words Laser microdissection, Laser capture, Postmortem tissue, Human brain, Proteomics, Mass spectrometry, Immunohistochemistry, Single cell

1 Introduction

The brain is a complex and heterogeneously built organ composed of different regions and layers, each with different combinations of cell types. On top of this, the etiology of neurodegenerative diseases is complicated in nature and they affect different regions and cell types in diverse manners. Information on differences in the proteome during development of a disease can provide valuable clues on which processes are involved. This can lead to identification of therapeutic targets and biomarkers for accurate and early diagnosis, monitoring of disease progression, and monitoring response to a therapeutic intervention. To obtain relevant information on the proteome on heterogeneous brain tissue, it is desirable to increase the spatial resolution of the analysis. For example, to focus the analysis on a specific region or cell type that is highly vulnerable in a disease or that expresses specific markers, which are relevant for diagnosis and prognosis.

Ka Wan Li (ed.), *Neuroproteomics*, Neuromethods, vol. 146,
https://doi.org/10.1007/978-1-4939-9662-9_4, © Springer Science+Business Media, LLC, part of Springer Nature 2019

Laser microdissection (LMD) is an ideal tool for both the rapid collection of large amounts of tissue and the pooling of larger numbers of single cells [1], and it has been applied for high-precision spatial-proteomics studies of the brain. For LMD followed by a proteomics analysis, frozen tissue sections are preferred [2, 3], but the more commonly available formalin-fixed paraffin-embedded tissues may also be used [4]. From the frozen tissue, thin sections are cut using a cryostat. The thickness is dependent on the size of the region of interest, which can vary between 5 and 50 μm. Tissue sections, applied on specific LMD-dedicated PEN foil slides, can be stained using histological techniques or by immunohistochemistry, allowing visualization of various morphological features or specific proteins or epitopes. The LMD microscope is equipped with a camera coupled to a computer, which facilitates a live view of the tissue on the screen. The software allows easy selection of the regions of interest by simply drawing them, using either the mouse or using a touchscreen. Regions of interest can be large, for example a specific region of the hippocampus [3], but it is also possible to isolate individual cells or even subcellular structures [5].

With the increased speed and sensitivity of current mass spectrometers it is becoming possible to quantify a large proportion of the proteome using only small amounts of input material. Thus, high sensitivity is becoming increasingly compatible with the analysis of relatively small amounts of tissue (protein) isolated using laser microdissection.

In this chapter we describe two methods of laser microdissection-assisted mass spectrometry analysis. The first describes the visualization of the CA1 and subiculum region of the brain's hippocampus using the histochemical toluidine blue staining and the subsequent isolation, sample preparation and LC-MS/MS analysis workflow as described previously [3] and referred to as protocol A. The second approach, referred to as protocol B, describes the workflow that includes the visualization of phospho-tau using immunohistochemistry, the isolation of phospho-tau positive pyramidal neurons, and subsequent sample preparation and LC-MS/MS analysis. For all microdissections a Leica LMD 6500 system was used. The collected sample can be processed by protocols based on the use of SDS-page gel as described in this chapter. Alternatively, Filter-aided sample preparation (FASP) or single-pot solid-phase sample preparation (SP3) protocols can be used (*see* Chapter 6). After digestion of the proteins, the resulting tryptic peptides can be analyzed by DDA, DIA (SWATH, *see* Chapter 11), or Isobaric multiplex labeling strategies (*see* Chapter 10) for relative quantitative proteomics.

With this chapter we aim to provide the tools to successfully perform a proteomics analysis that is focussed on small tissue regions and cell populations using (immuno-)histochemistry followed by a laser dissection-assisted LC-MS/MS workflow.

2 Materials

2.1 Apparatus

1. Laser dissection microscope. Here the LMD6500 system from Leica is used.
2. Cryostat (Leica).
3. Hair dryer.
4. Centrifuge, for example an Eppendorf 5415D system.
5. Heat block that is set at 95 °C.
6. SDS-PAGE electrophoresis system, for example an XCell SureLoc Mini-Cell Electrophoresis System from ThermoScientific and a Mini-PROTEAN® 3 Cell (Bio-rad).
7. 0.5 mL cap (Greiner Bio-one) or adhesive caps (Zeiss). Adhesive caps are obligatory when using a Zeiss PALM system but are useful in certain cases for the Leica system as well (*see* **Note 1**).

2.2 Reagents

2.2.1 Preparation of the Slides for Microdissection

1. Snap-frozen, postmortem, human brain tissue with short (<12 h) postmortem delay (*see* **Note 2**).
2. Polyethylene naphthalate (PEN)-foil membrane slides (Leica).
3. Ethanol 100%.
4. Ultrapure sterile H_2O.
5. Toluidine blue (Sigma-Aldrich) 1% (w/v) in sterile H_2O.
6. Sterile PBS: NaCl 8.2 g, $Na_2HPO_4 \cdot 12H_2O$ 3.1 g, $NaH_2PO_4 \cdot 2H_2O$ 0.3 g in 1000 mL (pH 7.4).
7. Primary antibody anti phospho-tau (clone AT8, Thermo scientific).
8. HRP-labeled secondary antibody Goat anti-mouse-HRP (DAKO).
9. 3,3′-Diaminobenzidine (DAB) in chromogen solution (DAKO).

2.2.2 SDS-PAGE

1. Pierce™ Lane Marker Reducing Sample Buffer 5× (Thermo Fisher Scientific).
2. M-PER lysis buffer (Thermo scientific).
3. Prestained protein ladder (Bio-Rad).
4. NuPAGE® 4-12% Bis-Tris acrylamide gel (Thermo Fisher Scientific).
5. NuPAGE™ MOPS SDS Running Buffer (20×) (Thermo Fisher Scientific).
6. 30% Acrylamide/Bis Solution, 19:1 (Bio-rad).
7. Tris–HCl 1.5 M, adjusted to pH 8.8 using HCl.

8. 10% w/v SDS in H_2O.

9. Ammonium persulfate: 10% (w/v) solution in H_2O. This can be stored at −20 °C until use, or prepare fresh solution just before use.

10. TEMED (Bio-rad).

11. 10× Tris/Glycine/SDS (Bio-rad).

12. Gel fixation solution: 50% (v/v) Ethanol and 3% (v/v) Phosphoric Acid (from 85% stock) in H_2O.

13. Colloidal Coomassie blue: 34% (v/v) Methanol, 3% (v/v) Phosphoric Acid (from 85% stock) (Sigma Aldrich), 150 g Ammonium Sulphate, 1 g Coomassie brilliant blue G-250 (Thermo Scientific) make up to 1 L in H_2O.

14. Ultrapure sterile H_2O.

3   Methods

3.1   Preparation of the Slides for Microdissection

1. Place PEN-foil slides in UV light for 30 min following the manufacturer's instructions.

2. Cut the tissue sections in a cryostat at a temperature of −16 °C to −18 °C, and transfer them onto the PEN-foil slides (*see* **Note 3**). In case large areas of tissues are required for analysis (protocol A), the cut section thickness should be between 10 and 50 μm depending on the tissue type and the type of laser dissection microscope (*see* **Note 4**). For the isolation of individual cells (protocol B) the thickness used is typically 10 μm or less (*see* **Note 5**).

3. Air-dry the tissue sections for at least 10 min.

4. Fix the sections in ethanol for 1 min (*see* **Notes 6** and **7**), and air-dry for 10 min. Alternatively it can be dried in 1 min when a hair dryer is used, which is set to cool air.

5. Proceed with Subheading 3.1.1 or 3.1.2.

3.1.1   Preparing Sections with Toluidine Blue Staining (Protocol A)

1. Briefly wet the sections with water, remove excess water by tilting the slide.

2. Apply a few drops of toluidine blue solution to completely cover the tissue, and incubate for 1 min.

3. Wash the slides for 30 s in a container containing 250 mL or more water.

4. Repeat the washing once more in another container containing 250 mL water.

5. Wash the slides in a container with 250 mL or more ethanol.

6. Repeat the washing twice sequentially in two other containers with 250 mL ethanol.

7. Air-dry the slides for 10 min. Alternatively, dry the slides with a hair dryer that is set to cool air for 1 min.

8. Store at room temperature till further use. To preserve the quality the slides should be used within 1 week (*see* **Note 8**). A typical image obtained using this protocol is shown in Fig. 1a.

3.1.2 Fast Immuno-Histochemical Staining (Protocol B)

1. Prepare antibody dilutions in 100 μL PBS per section (*see* **Note 9**).

2. Briefly wet the sections with PBS, and remove excess fluid.

3. Apply the diluted primary antibody to the section. Make sure the entire tissue section is covered by distributing the liquid with the back of a pipet tip.

4. Incubate for 20 min.

5. Wash the slide for 30 s in a container with 250 mL or more of PBS.

6. Repeat the washing twice in two other containers with 250 mL of PBS.

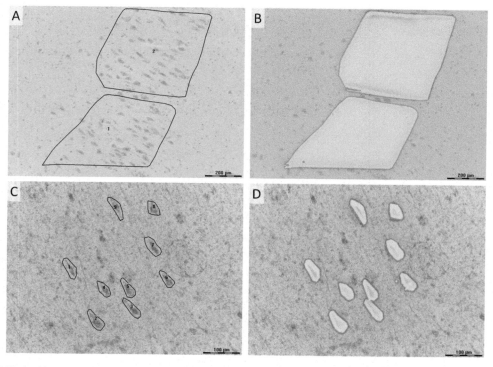

Fig. 1 Typical images obtained using the different (I)HC protocols before and after tissue isolation. (**a**) Toluidine blue staining of human hippocampus. Shown is the CA1 region of the hippocampus. Two areas were selected for dissection. (**b**) The same area after laser dissection. (**c**) Cells displaying immunoreactivity for phospho-tau visualized using DAB (brown), toluidine blue was used as a counterstain (blue). Several phospho-tau positive cells are selected for dissection. (**d**) Same area after laser dissection. Scale bar indicates 200 μm (**a, b**) or 100 μm (**c, d**)

7. Apply the secondary HRP-labeled antibody at an appropriate dilution, depending on the antibody used.

8. Incubate for 20 min.

9. Wash three times 30 s in PBS.

10. Prepare DAB solution following the manufacturer's protocol, and add to the slide.

11. Incubate for 5 min in the dark.

12. Wash the slides thoroughly in water.

13. Incubate with toluidine blue in water for 1 min as a counterstain.

14. Wash the slides twice for 30 s in water.

15. Wash the slides in ethanol three times for 30 s.

16. Air-dry for 10 min, or dry using a hair dryer, set to cool air, for 1 min.

17. Store at room temperature till further use, preferable within 1 week (*see* **Note 8**). A typical image using this type of protocol is shown in Fig. 1c.

3.2 Laser Dissection Procedure

1. Turn the microscope and the computer on:

   (a) Turn the microscope controller (Leica CTR 6500) and the computer on.

   (b) Turn on the CryLas (Laser) by turning the key clockwise.

   (c) Start the program: Leica Laser Microdissection.

2. Open the "Laser Control" and "Microscope Control" menus (if not already open).

   (a) Tab "Laser"—"Control".

   (b) Tab "Options"—"Microscope Control".

3. Press the icon "Unload slide" and place the PEN-foil slide holding the tissue section with the tissue facing down.

4. Press the icon "Unload collector" for the collector and insert Eppendorf tube(s).

   (a) Depending on the availability and amount of material that has to be dissected, either load the cap with 30 μL of 1× reducing SDS loading buffer (diluted with M-PER lysis buffer), or place an adhesive cap (Zeiss) in the cap holder below the specimen (*see* **Note 10**).

5. Proceed with dissection of either large areas (protocol A, Subheading 3.2.1) or small areas, like individual cells (protocol B, Subheading 3.2.2).

3.2.1 Laser Dissection Procedure (Protocol A)

1. To obtain approximately 40 µg of protein, a volume of 1×10^9 µm³ has to be isolated. This is best to be divided over two separate Eppendorf tubes (60 µL SDS loading buffer in total).

2. Large areas are dissected most efficiently using either a 5× or 10× objective.

3. Start with Laser Calibration, choose the tab "Laser"—"Calibrate".

4. Laser Control settings (Power, Aperture, Speed, etc.) need to be optimized by the user, generating a thin cutting line.

5. Select area(s) of interest, dissect and collect the tissue using the LDM system according the manufacturer's instructions. An example picture of the tissue before and after laser dissection is shown in Fig. 1a, b.

6. After tissue isolation freeze at −80 °C until further use.

3.2.2 Laser Dissection Procedure (Protocol B)

1. Small areas like individual cells are dissected most efficiently using either a 20× or 40× objective.

2. Start with Laser Calibration, choose tab Laser-Calibrate.

3. Laser Control settings (Power, Aperture, Speed, etc.) need to be optimized by the user, generating a thin cutting line.

4. Start each new slide with making a small hole in the PEN membrane at a location without tissue to let the air out (*see* **Note 11**).

5. Select area(s) of interest, dissect and collect the tissue using the LDM system according the manufacturer's instructions. An example picture of the tissue before and after laser dissection is shown in Fig. 1c, d.

6. When the isolated material is in reducing SDS sample buffer, store in −80°C until further use.

7. When the isolated material is in an adhesive cap, then proceed with taking it up in reducing SDS sample buffer as follows:

    (a) Add 25 µL of reducing SDS sample buffer.

    (b) Incubate for 10 min at RT.

    (c) Pipet "up and down" repeatedly (5–10 times).

    (d) Transfer to a new Eppendorf tube.

    (e) Check the adhesive cap under a microscope to convince yourself that all material is removed from the adhesive cap.

8. Store the sample at −80 °C until further use.

3.3 SDS-PAGE

3.3.1 SDS-PAGE for Large Areas of Interest (Protocol A)

1. Let the sample defrost.
2. Heat the samples to 95 °C and incubate for 5 min.
3. Let the samples cool down to RT.
4. -Optional- incubate with 50 mM iodoacetamide for 30 min at RT in the dark.
5. Separate proteins based on their size on a NuPAGE® 4–12% Bis-Tris acrylamide gel (Invitrogen) using MOPS SDS running buffer (Invitrogen) according the manufacturer's protocol.
6. After electrophoresis, remove the gel from the cassette.
7. Incubate the gel in gel fixing solution for at least 2 h.
8. Wash the gel in ultrapure sterile water for 10 min.
9. Incubate in Coomassie blue solution for 10 min.
10. Wash the gel in ultrapure sterile H_2O for several hours, refreshing the H_2O repeatedly, until the gel is de-stained.
11. The gel can now be cut into the required number of fractions, which depends on the type of mass spectrometer used. For the LTQ-Orbitrap mass spectrometer we used 12 fractions. For Triple TOF 5600 system we used 2–4 fractions. For a Fusion Lumos Orbitrap mass spectrometer it is possible to quantify 4000 proteins from a single fraction using a 2–3 h HPLC gradient (*see* Subheading 3.3.2 for SDS-PAGE of proteins in a single fraction).
12. Proceed with in gel trypsin digestion protocol.

3.3.2 SDS-PAGE for Individually Isolated Cells (Protocol B)

1. Make a 10% gel, of 1 mm thickness with 10 wells as follows (for 10 mL):

   (a) Mix 3.3 mL of the 30% acrylamide, 1.5 mL 1.5 M Tris (pH 8.8) and 4.96 mL ultrapure H_2O.

   (b) Add APS 0.1 mL and TEMED 0.006 mL, mix gently, and directly pipette this into the gel cast as polymerization starts directly. Place the 10-well comb and wait for the gel to polymerize (typically <30 min).

2. Place the prepared gel in the electrophoresis system and fill up with TGS running buffer 1×.
3. Remove the comb.
4. Load the samples into the gel wells.
5. Let the proteins run into the gel by electrophoresis (150 V) until the sample is in the gel for a length of approximately 8–10 mm.
6. Remove the gel from the cast.
7. Incubate the gel in gel fixing solution for at least 2 h.

8. Wash the gel in ultrapure sterile water for 10 min.

9. Incubate in Coomassie blue solution for 10 min or longer if desired (*see* **Note 12**).

13. Wash the gel in ultrapure sterile H_2O for several hours, refreshing the H_2O repeatedly, until the gel is de-stained.

14. Cut out the gel piece containing the sample and proceed with in gel digestion. The gel piece can be divided into multiple pieces if desired.

15. Proceed with in gel trypsin digestion protocol.

3.4 In Gel Trypsin Digestion and Mass Spectrometry

Digest proteins contained in the gel by trypsin as detailed in Chapters 8 and 11. The resulting tryptic peptides can be analyzed by any modern LC-MS/MS system. For general description *see* Chapter 14; for data-dependent acquisition (DDA) *see* Chapters 5, 8, and 9; for data-independent acquisition (DIA, also called SWATH) *see* Chapter 11; for Isobaric multiplex labeling strategies for relative quantitative proteomics (TMT) *see* Chapter 10.

4 Notes

1. Adhesive caps can also be used on a Leica system. This can be of practical use, especially when isolating individual cells of a type that is low abundant and therefore requires multiple days of laser dissection. The samples are then collected without buffer and can therefore be stored dry, preventing degradation.

2. Rapid freezing of tissue using liquid nitrogen is highly important. When the freezing procedure is too slow, this will destroy the morphology and making the tissue crumble when cutting the tissue, especially at larger thickness.

3. Check if the PEN-membrane is intact. When the membrane is damaged then fluid will get between the foil and the glass, rendering the slide unusable for LMD. This is especially relevant when isolating larger areas of tissue (protocol A).

4. For analysis of larger areas of interest, usually thicker sections are preferred as it saves time in performing laser dissection and reduces the amount of (expensive) PEN-foil slides needed. Modern laser dissection microscopes, like the Leica LMD 6500, have sufficient laser power to cut through sections up to 50 μm.

5. For the isolation of separate cells or other very small structures, sections are preferably thin, 10 μm or less. Thickness depends largely on the size of the objects that you wish to isolate, considering that when the section is thicker than the object of

interest then tissue surrounding the object will be isolated as well, reducing the purity of the sample.

6. Typically we fix tissue in 100% ethanol; however, we experienced that some IHC stainings were incompatible with ethanol fixation, but were successful when fixing using 100% acetone. No negative effects were observed in the downstream LC-MS/MS analysis.

7. Be careful not to damage the PEN-membrane slides during the fixation or staining procedure as they can become unusable (also *see* **Note 3**).

8. Proteins in the tissue are, when dry, very resistant to degradation. It is often preferred to store the slides at RT instead of freezing them. Freezing introduces the risk of deposition of moisture, and freeze thaw cycles are best avoided. However, fast (within a week) processing of the sections is recommend; meaning finishing the laser dissection and storing these in reducing SDS buffer at −80 °C.

9. Usually the antibody concentration used for a fast IHC staining is about ten fold higher as used in a normal overnight incubation for IHC. However performing a test series with different antibody concentrations is advised prior to the actual experiment. Antibodies that produce a highly contrasted and specific staining will give the best chance of success, as the fast IHC protocol will produce a staining of lower quality than an overnight incubation.

10. To decide for either the cap with reducing SDS sample buffer or an adhesive cap is particularly valid when isolating separate cells (protocol B). For our experiments we found a number of 3000 cells to provide satisfactory results. Although even higher numbers will evidently result in more proteins that can be quantified as well as provide a more reliable quantification. When a particular cell or structure of interest is very rare in the tissue, it can take several days to isolate sufficient amount of cells for a successful MS analysis. Our experience is that it is then useful to capture the material in adhesive caps. The material is fairly well fixed in place and you can store it at RT and easily continue the next day with dissection using the same cap.

11. Usually dissection of small structures like individual cells, inclusion bodies, or protein aggregates is done at high magnification, for example, using the 20× or the 40× objective. The PEN-membrane glass slide usually has some air between the glass and the foil. When selecting a number of shapes to be dissected you will find that when the first hole is made in the foil, the air will come out and the foil will move towards the glass. Consequentially, your tissue section will be out of focus

for all remaining shapes, which will compromise the precision and effectivity of the dissection. This step is not required when metal frame slides are used.

12. Since only very little amount of protein is isolated, the Coomassie staining will be very weak. Longer incubation time can increase the signal. However, complete lack of staining does not necessarily indicate absence of proteins. Successful quantification of a high number of proteins can still be realized using a sensitive mass spectrometer.

References

1. Datta S, Malhotra L, Dickerson R, Chaffee S, Sen CK, Roy S (2015) Laser capture microdissection: Big data from small samples. Histol Histopathol 30:1255–1269

2. Hondius DC, Eigenhuis KN, Morrema THJ et al (2018) Proteomics analysis identifies new markers associated with capillary cerebral amyloid angiopathy in Alzheimer's disease. Acta Neuropathol Commun 6:46

3. Hondius DC, van Nierop P, Li KW, Hoozemans JJ, van der Schors RC, van Haastert ES, van der Vies SM, Rozemuller AJ, Smit AB (2016) Profiling the human hippocampal proteome at all pathologic stages of Alzheimer's disease. Alzheimers Dement 12:654–668

4. Drummond E, Nayak S, Faustin A et al (2017) Proteomic differences in amyloid plaques in rapidly progressive and sporadic Alzheimer's disease. Acta Neuropathol 133:933–954

5. Wong TH, Chiu WZ, Breedveld GJ et al (2014) PRKAR1B mutation associated with a new neurodegenerative disorder with unique pathology. Brain 137:1361–1373

## Chapter 5

High-pH Reversed-Phase Pre-Fractionation for In-Depth Shotgun Proteomics

Ning Chen, Mingwei Liu, Jun Qin, Wei Sun, and Fuchu He

Abstract

Two-dimensional LC-MS-based proteomics is a powerful strategy for high-throughput protein identification and quantification. The combination of high-pH reversed-phase fractionation with low-pH LC-MS analysis is the method of choice, because it provides the optimal separation efficiency and deep protein coverage. In this chapter, we illustrate the protocols of HPLC-based and the tip-based high-pH reversed-phase chromatographic fractionation of peptide mixtures, in which the HPLC-based reversed-phase fractionation provides higher resolving power while the tip-based high-pH reversed-phase fractionation is simple, parallelizable, and suitable for limited amount of biological samples.

Key words High pH, Reverse phase, Fractionation, Liquid chromatography, Two-dimensional chromatography, Mass spectrometry, Proteomics

1 Introduction

Biological samples are complex and often comprise tens of thousands of proteins. Enzymatic digestion by proteases, such as trypsin, further produces innumerable peptides. A single-step analysis of liquid chromatography (LC) coupled to mass spectrometer may not have enough analytical power to resolve and identify this high number of peptides. In particular, many of the co-eluted peptides may have similar mass-to-charge (m/z) ratios and different levels of abundance. Therefore, for in-depth quantitative proteomics an extra fractionation step is often implemented. Currently, two-dimensional chromatography coupled to mass spectrometry has rapidly grown in use for the comprehensive analysis of peptide constituents [1–4].

Mainstream two-dimensional fractionation techniques consist of two independent chromatographic techniques based on hydrophobicity, charge, molecular weight, or affinity of peptides [2, 5]. In most shotgun proteomic analyses, the second dimension is

---

Ning Chen and Mingwei Liu contributed equally to this work and are co-first authors.

Ka Wan Li (ed.), *Neuroproteomics*, Neuromethods, vol. 146,
https://doi.org/10.1007/978-1-4939-9662-9_5, © Springer Science+Business Media, LLC, part of Springer Nature 2019

performed by low-pH reversed-phase liquid chromatography (RPLC) because of its superior separation efficiency and excellent compatibility with the mass spectrometer [6]. The fractionation of peptide mixtures by two-dimensional LC techniques has been performed using various orthogonal combinations, such as high-pH reversed phase/low-pH reversed-phase liquid chromatography (high-pH RP/low-pH RPLC), strong cation exchange/reversed-phase liquid chromatography (SCX/RPLC), size exclusion chromatography/reversed-phase liquid chromatography (SEC/RPLC), and affinity chromatography/reversed phase liquid chromatography (AFC/RPLC) [2, 3]. By combining the resolving power of two independent chromatography modes, the complex peptide mixtures can be further discriminated due to the higher separation efficiency and greater peak capacity.

RPLC fractionation of peptide mixtures is based on hydrophobic interactions of the peptides with a hydrophobic stationary phase such as octadecyl (C_{18}) and a polar hydrophilic mobile phase [7]. High-pH RPLC was introduced by Gilar et al. as the first dimension of peptide mixtures fractionation showing great orthogonality with low-pH RPLC [8, 9]. The significant difference in chromatographic selectivity between high-pH RP and low-pH RP is due to the dramatic change in charge distribution within the peptide chain. Since peptides are charged molecules comprised of ionizable basic and acidic functional groups, their net charges are affected by the mobile phase pH. The acidic peptides are retained longer at low-pH due to the increased hydrophobicity of protonated carboxylic residues, while the basic peptides are retained longer at high pH due to the increased hydrophobicity of deprotonated basic residues [2, 8]. Comparing with SCX/RPLC and SEC/RPLC, high-pH RP/low-pH RPLC followed by fraction concatenation provides the optimal resolution and higher proteome coverage for shotgun proteomic analysis [3, 9, 10]. Furthermore, high-pH RPLC generates cleaner fractions for downstream LC-MSMS analysis due to the use of salt-free buffers, whereas desalting is necessary for SCX fractions [11]. It can decrease sample loss and reduce processing time.

Currently, the high-pH RP/low-pH RPLC strategy has been employed in various proteomic applications to separate isotopic-labeled [12, 13] or non-labeled peptides [14, 15]. In this chapter, two methods of high-pH RPLC are illustrated, including the traditional HPLC-based separation and the tip-based chromatographic separation. The HPLC-based high-pH RP separation followed by fraction concatenation provides greater resolving power [11], whereas the tip-based high-pH RP separation can be applied in parallel and suitable for limited amount of biological samples [16, 17]. The readers can choose the high-pH RP separation strategy according to the facility available in lab, analysis depth, sample number, and protein amount.

2 Materials

All solvents and buffers are prepared with ultrapure water and chemicals with high purity.

2.1 Sample Preparation, Lysis, and Digestion

For materials related to brain tissues or synaptic fractions isolation, please refer to Chapters 2–4. For the materials related to lysis and digestion, please refer to Chapters 6 and 11.

2.2 HPLC-Based High-pH RP Fractionation and Concatenation

1. High-pH compatible C_{18} RP column, 130 Å, 3.5 μm, 2.1 mm × 150 mm (XBridge Peptide BEH C_{18} Column, stable at pH 1.0–12.0, Cat. No. 186003565, Waters).

2. HPLC system with following requirements: flow rates ranging from 0.1 to 4 mL/min, 0.1 to 1 mL sample loop; UV detector set to fixed wavelengths of 214 nm; compatible up to 4000 psi; stable at pH 2.0–12.0; auto fraction collector is recommended.

3. High-pH mobile phase A: 2% acetonitrile (ACN), adjusted pH to 10.0 using $NH_3 \cdot H_2O$.

4. High-pH mobile phase B: 98% ACN, adjusted pH to 10.0 using $NH_3 \cdot H_2O$.

2.3 Tip-Based High-pH RP Fractionation and Concatenation

1. Empore C_{18} (Octadecyl) Solid Phase Extraction Disks (Cat. No. 2215, 3 M Purification).

2. High-pH compatible C_{18} beads (Durashell C_{18} (L) beads, 150 Å, 3 μm, stable at pH 1.5–12.0, Cat. No. DC930001-L, Agela Technologies).

3. 16-Gauge blunt-tipped needle.

4. 200-μL plastic pipette tips.

5. Centrifuge adapters (Cat. No. 5010-21514, GL Sciences).

6. 1.5 mL Eppendorf tubes.

7. 10 mM NH_4HCO_3, adjusted pH to 10.0 using $NH_3 \cdot H_2O$.

8. High-pH RP elution solvents: Mix ACN and 10 mM NH_4HCO_3 (pH 10.0) to obtain the elution solvents containing 6%, 9%, 12%, 15%, 18%, 21%, 25%, 30% and 35% ACN (pH 10.0).

2.4 Nano-LC/MSMS and Data Analysis

1. Nano-LC mobile phase A: 0.1% formic acid, 2% ACN.

2. Nano-LC mobile phase B: 0.1% formic acid, 98% ACN.

3. C_{18} RP column: Capillary column (150 μm × 120 mm) packed with C_{18} beads (120 Å, 1.9 μm).

4. Nano-LC/MSMS: nano-LC coupled with high performance mass spectrometer, such as Orbitrap HF (Thermo Fisher).

3 Methods

3.1 Sample Preparation, Lysis, and Digestion

For methods related to brain tissues or synaptic fractions isolation, please refer to Chapters 2–4. For methods related to lysis and digestion, please refer to Chapters 6 and 11. Cell lysis additives such as detergents and reducing agents that may interfere with fractionation or other downstream analysis must be removed. We recommend using at least 50 µg protein digest for the chromatographic fractionation with high-pH compatible C_{18} RP column (130 Å, 3.5 µm, 2.1 mm × 150 mm), while using at least 10 µg protein digest for tip-based chromatographic fractionation.

3.2 HPLC-Based High-pH RP Fractionation and Concatenation

1. HPLC system preparation: Connect the column to HPLC system, prime and equilibrate the column with mobile phases. The flow rate is determined according to the column internal diameter and manufacturer instructions. Column pressure and UV absorbance at 214 nm are monitored overtime (*see* **Note 1**). The column is first primed by running mobile phase B for at least 40 min, and then washed with mobile phase A for at least 30 min. For each step, maintain the flow rate on the column until stable UV baseline and expected pressure are achieved.

2. Sample loading: Dissolve the dried peptides in 100 µL mobile phase A. Vortex and spin at the maximum speed for 10 min. Take the supernatant and inject it slowly into the column (*see* **Note 2**).

3. Chromatographic fractionation: Peptides bound to the column are eluted sequentially with the increasing concentration of ACN. The mobile phase gradient is set as follows: 5% B, 5 min; 5–18% B, 45 min; 18–35% B, 30 min; 35–95% B, 3 min; 95% B, 10 min; 95–5% B, 3 min; 5% B, 2 min (*see* **Note 3**). Eluent is collected every 100 s (Fig. 1) (*see* **Note 4**).

4. Fraction concatenation: Multiple early, middle, and late RPLC fractions eluted over equal time intervals are pooled into a single concatenated subfraction (*see* **Note 5**). In case to concatenate to 8 subfractions, fractions 1, 9, 17, 25 33, and 41 are combined as subfraction 1; fractions 2, 10, 18, 26, 34 and 42 are combined as subfraction 2; et al. (*see* **Note 6**).

5. Drying and storage: Dry the subfractions using a vacuum centrifuge or lyophilizer, and store them at −20 °C until use.

6. Postfractionation column care: After the high-pH RP fractionation, it is recommended to run a post wash with buffers containing only water and ACN without ammonia. The wash gradient is set as follows: 5–90% ACN, 15 min; 90% ACN, 10 min; 90–5% ACN, 5 min; 5% ACN, 5 min; 5–90% ACN, 15 min; 90% ACN, 10 min; 90–5% ACN, 5 min; 5% ACN, 5 min. Then increase the ACN concentration up to 80% and

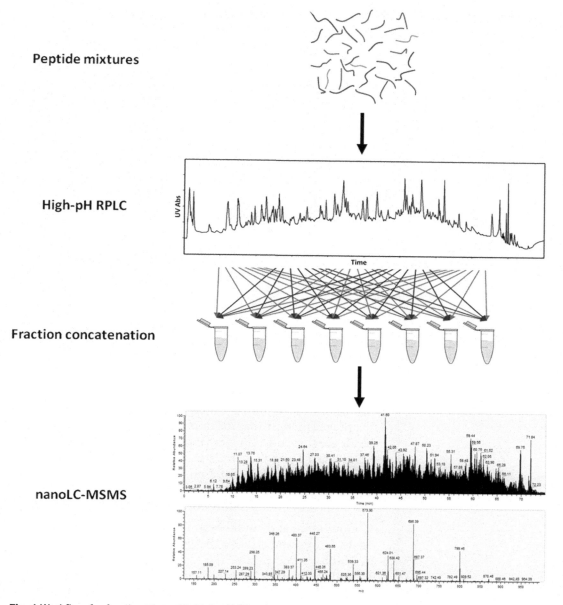

Fig. 1 Workflow for fractionation with high-pH RP column

keep on running for 20 min. Reduce the flow rates step by step until the pressure is diminished. Remove the column, cap both sides, and store it at 4 °C.

3.3   Tip-Based High-pH RP Fractionation and Concatenation

In the following steps of C_{18} tip conditioning and equilibration, sample loading and fractionation, centrifugation is performed at $800 \times g$ until the last remainder of buffer just covers the beads.

1. C_{18} tip construction: A small disk of C_{18} membrane is stamped out from Empore material using a 16-gauge blunt-tipped

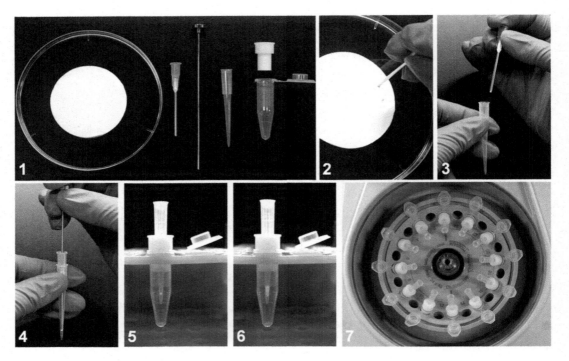

Fig. 2 RP (C$_{18}$) tip preparation

needle, and then plug the membrane into a 200-µL tip. Place the StageTip with adaptor into a 1.5 mL eppendorf tube. C$_{18}$ beads are suspended in methanol and then transferred into the StageTips (2 mg C$_{18}$ beads per tip (*see* **Note 7**)). After centrifugation at $1000 \times g$ for 3 min, C$_{18}$ beads are loosely packed at the bottom of the tip (Fig. 2). The packed C$_{18}$ tips can be stored at room temperature for a long time.

2. Conditioning and equilibration: Add 200 µL ACN to the top of the C$_{18}$ tip and remove the air bubbles. After centrifugation at $1000 \times g$ for 3 min, the C$_{18}$ tip is washed with high-pH solvents containing 50%, 80% and 50% ACN sequentially, and then equilibrated by washing with 10 mM NH$_4$HCO$_3$ (pH 10.0) twice. Each wash takes 100 µL solvent.

3. Sample loading: Dissolve the dried peptides in 10 mM NH$_4$HCO$_3$ (pH 10.0). Vortex and spin at the maximum speed for 10 min. Take the supernatant and load it slowly onto the C$_{18}$ tip. After centrifugation at $1000 \times g$ for 3–5 min, the flow through is reloaded on the C$_{18}$ tip and washed with 100 µL 10 mM NH$_4$HCO$_3$ (pH 10.0) twice.

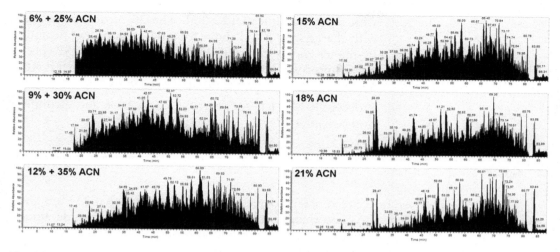

Fig. 3 Total ion current chromatograms displaying peptide elution patterns across the fractions from high-pH RP tip for proteomic analysis of mouse tissue

4. Chromatographic fractionation and combination: Peptides bound to the column are eluted with high-pH solvents containing 6%, 9%, 12%, 15%, 18%, 21%, 25%, 30%, and 35% ACN (pH 10.0) sequentially. Each elution takes 100 μL solvent. Combine the 6% ACN fraction with the 25% ACN fraction, and the 9% ACN fraction with the 30% ACN fraction, 12% ACN fraction with the 35% ACN fraction, but leave the 15%, 18%, and 21% ACN fractions separately (Fig. 3) (*see* **Note 8**).

5. Drying and storage: Dry the fractions using a vacuum centrifuge or lyophilizer and store them at −20 °C until use.

3.4 Nano-LC/MSMS and Data Analysis

1. Sample reconstitution: Redissolve the fractions/subfractions from the first dimension RPLC in 10 μL 0.1% formic acid. After centrifugation at the maximum speed for 10 min, 9 μL supernatant of each sample is transferred to the sample vial.

2. Nano-LC separation: 8 μL sample is loaded into the injection loop. Peptides are first trapped by a short trap column and then separated by an in-house-made C_{18} RP column using a 81 min linear gradient at a flow rate of 300 nL/min (5–11% B, 14 min; 11–23% B, 37 min, 23–36% B; 17 min; 36–100% B, 1 min; 100% B, 6 min).

3. Mass spectrometry detection: During the elution, peptides are directly injected into the Orbitrap HF mass spectrometer (Thermo Fisher) via electrospray ionization. For full MS survey scan, automatic gain control (AGC) target is 3×10^6, max injection time is 80 ms, and scan range is from 350 to 1500

with the resolution of 120,000. The 20 most intense peaks with charge state 2 and above were selected for fragmentation by higher-energy collision dissociation (HCD). MS2 spectra are acquired with 15,000 resolution. MS2 AGC target is set to 5.0×10^4 with a max injection time of 19 ms.

4. Process the MS data for protein identification and quantification (Chapter 12).

4 Notes

1. We recommend detecting the eluted peptides at 214 nm, since UV absorbance at 214 nm is specific for peptide bonds. Absorbance at 260 nm or 280 nm may also be used but not so accurate.

2. The injection volume is based on sample concentration, column capacity, and the loop volume.

3. Fundamentally, the more fractions that are concatenated from first-dimension separation, the more efficient use of a wider elution window in the second-dimension separation [11]. Besides, the overlap between neighboring fractions from the first-dimensional separation may result in the overlap of the neighboring concatenated fractions prepared for the second-dimensional separation [11, 18]. Therefore, we recommend a longer gradient with eluent collection at least minutes apart in the first dimension.

4. Since the fractions at the beginning (0–5% B) and the end (32–95–5% B) may contain minimal peptides but large amount of salts or chemicals, it is recommended not to take them for concatenation and downstream analysis.

5. Concatenation of high-pH RP fractions contributes higher efficient utilization of the separation window of the second dimension chromatography. Furthermore, it effectively reduces the analysis time required for the second dimension separation.

6. As an example of HPLC-based high-pH RP fractionation, similar number of unique peptides/proteins from human cells was identified across the concatenated fractions (Fig. 4a). More than 80% unique peptides were identified in only one fraction, indicating the great separation efficiency and the minor overlap between neighboring fractions (Fig. 4b). In addition, more than 90% proteins were detected with multiple unique peptides, suggesting the deep protein coverage (Fig. 4c).

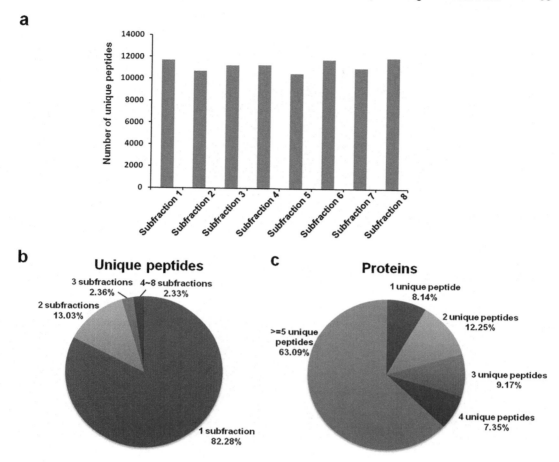

Fig. 4 Performance of HPLC-based high-pH RP fractionation for proteomic analysis of human cells. (**a**) The distribution of unique peptides among the high-pH RP fractions. (**b**) The percentage of unique peptides found in one or more fractions. (**c**) The percentage of proteins identified with one or more unique peptides

7. The amount of C_{18} beads for tip packing is determined according to the peptide amount for fractionation. Normally, 2 mg C_{18} beads are suitable for 10–30 μg peptides, while more C_{18} beads should be packed for larger amount of peptides.

8. As an example of tip-based high-pH RP fractionation, around 8500–13,500 unique peptides were identified across the RP fractions from mouse tissue (Fig. 5a). More than 60% unique peptides were identified in only one fraction (Fig. 5b), and more than 80% proteins were detected with multiple unique peptides (Fig. 5c). Tip-based high-pH RP fractionation followed by low-pH RPLC/MSMS is suitable for in-depth proteomic analysis of minor amount of biological samples.

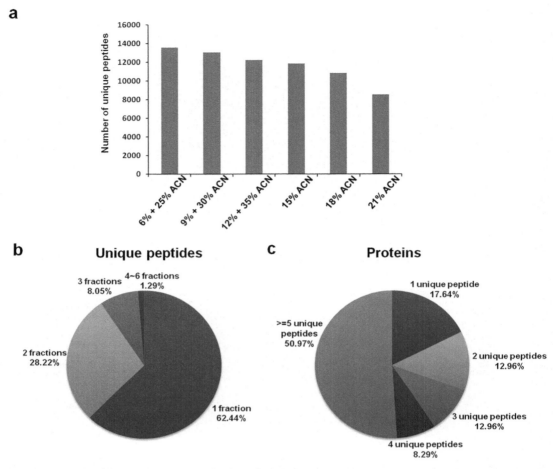

Fig. 5 Performance of tip-based high-pH RP fractionation for proteomic analysis of mouse tissue. (**a**) The distribution of unique peptides among the high-pH RP fractions. (**b**) The percentage of unique peptides found in one or more fractions. (**c**) The percentage of proteins identified with one or more unique peptides

References

1. Rappsilber J, Mann M, Ishihama Y (2007) Protocol for micro-purification, enrichment, pre-fractionation and storage of peptides for proteomics using StageTips. Nat Protoc 2:1896–1906

2. Di Palma S, Hennrich ML, Heck AJ, Mohammed S (2012) Recent advances in peptide separation by multidimensional liquid chromatography for proteome analysis. J Proteome 75:3791–3813

3. Horvatovich P, Hoekman B, Govorukhina N, Bischoff R (2010) Multidimensional chromatography coupled to mass spectrometry in analysing complex proteomics samples. J Sep Sci 33:1421–1437

4. Lee H, Mun DG, So JE, Bae J, Kim H, Masselon C, Lee SW (2016) Efficient exploitation of separation space in two-dimensional liquid chromatography system for comprehensive and efficient proteomic analyses. Anal Chem 88:11734–11741

5. Sandra K, Moshir M, D'Hondt F, Tuytten R, Verleysen K, Kas K, Francois I, Sandra P (2009) Highly efficient peptide separations in proteomics. Part 2: bi- and multidimensional liquid-based separation techniques. J Chromatogr B Analyt Technol Biomed Life Sci 877: 1019–1039

6. Sandra K, Moshir M, D'Hondt F, Verleysen K, Kas K, Sandra P (2008) Highly efficient peptide separations in proteomics part 1.

Unidimensional high performance liquid chromatography. J Chromatogr B Analyt Technol Biomed Life Sci 866:48–63

7. Mitulovic G, Mechtler K (2006) HPLC techniques for proteomics analysis--a short overview of latest developments. Brief Funct Genomic Proteomic 5:249–260

8. Gilar M, Olivova P, Daly AE, Gebler JC (2005) Two-dimensional separation of peptides using RP-RP-HPLC system with different pH in first and second separation dimensions. J Sep Sci 28:1694–1703

9. Gilar M, Olivova P, Daly AE, Gebler JC (2005) Orthogonality of separation in two-dimensional liquid chromatography. Anal Chem 77:6426–6434

10. Dowell JA, Frost DC, Zhang J, Li L (2008) Comparison of two-dimensional fractionation techniques for shotgun proteomics. Anal Chem 80:6715–6723

11. Yang F, Shen Y, Camp DG 2nd, Smith RD (2012) High-pH reversed-phase chromatography with fraction concatenation for 2D proteomic analysis. Expert Rev Proteomics 9:129–134

12. Chen Z, Wen B, Wang Q, Tong W, Guo J, Bai X, Zhao J, Sun Y, Tang Q, Lin Z, Lin L, Liu S (2013) Quantitative proteomics reveals the temperature-dependent proteins encoded by a series of cluster genes in thermoanaerobacter tengcongensis. Mol Cell Proteomics 12:2266–2277

13. O'Brien DP, Timms JF (2014) Employing TMT quantification in a shotgun-MS platform. Methods Mol Biol 1156:187–199

14. Ding C, Jiang J, Wei J, Liu W, Zhang W, Liu M, Fu T, Lu T, Song L, Ying W, Chang C, Zhang Y, Ma J, Wei L, Malovannaya A, Jia L, Zhen B, Wang Y, He F, Qian X, Qin J (2013) A fast workflow for identification and quantification of proteomes. Mol Cell Proteomics 12:2370–2380

15. Farrell A, Mittermayr S, Morrissey B, Mc Loughlin N, Navas Iglesias N, Marison IW, Bones J (2015) Quantitative host cell protein analysis using two dimensional data independent LC-MS(E). Anal Chem 87:9186–9193

16. Chen W, Wang S, Adhikari S, Deng Z, Wang L, Chen L, Ke M, Yang P, Tian R (2016) Simple and integrated Spintip-based technology applied for deep proteome profiling. Anal Chem 88:4864–4871

17. Dimayacyac-Esleta BR, Tsai CF, Kitata RB, Lin PY, Choong WK, Lin TD, Wang YT, Weng SH, Yang PC, Arco SD, Sung TY, Chen YJ (2015) Rapid high-pH reverse phase StageTip for sensitive small-scale membrane proteomic profiling. Anal Chem 87:12016–12023

18. Wang H, Sun S, Zhang Y, Chen S, Liu P, Liu B (2015) An off-line high pH reversed-phase fractionation and nano-liquid chromatography-mass spectrometry method for global proteomic profiling of cell lines. J Chromatogr B Analyt Technol Biomed Life Sci 974:90–95

## Chapter 6

Sample Preparation for Proteomics Analysis: Filter-Aided Sample Preparation (FASP) and Single-Pot Solid-Phase Sample Preparation (SP3)

Ka Wan Li

Abstract

The success of a proteomics experiment critically depends on the correct execution of the sample preparation. Gel-based sample preparation is a well-proven method widely used in qualitative and quantitative proteomics. Recently, sample preparation in a single vessel has gained popularity because of the ease to perform, the possibility for large-scale analysis, and the potential to improve sensitivity. In this chapter I describe two sample preparation protocols, filter-aided sample preparation (FASP) and single-pot solid-phase sample preparation (SP3).

Key words Sample preparation, FASP, SP3, Proteomics

1 Introduction

Bottom-up proteomics is the method of choice for the identification and quantification of thousands of proteins from a biological tissue. The depth of proteome analysis and the speed of measurement are influenced primarily by the hardware configuration, for example, the type of mass spectrometer. The overall success of the experiment, however, is critically dependent on the use of an optimal condition for sample treatment upstream of LC-MS/MS analysis. A sample preparation protocol for proteomics analysis should effectively extract proteins from the tissue of interest in a condition that is compatible with subsequent protease digestion, and the resulting peptides should be quantitatively recovered for LC-MS/MS analysis.

Typically, the biological tissue is homogenized in a buffer containing denaturing reagents to solubilize the maximum amount of proteins (including the hydrophobic membrane proteins), which are then digested by trypsin (in some cases in combination with EndoLys-C). The most commonly used denaturing reagents

Ka Wan Li (ed.), *Neuroproteomics*, Neuromethods, vol. 146,
https://doi.org/10.1007/978-1-4939-9662-9_6, © Springer Science+Business Media, LLC, part of Springer Nature 2019

include chaotropic chemicals (e.g., urea and guanidine hydrochloride) and ionic detergents (e.g., sodium dodecyl sulfate). However, these reagents are known to negatively affect trypsin activity, and may not be compatible with LC-MS/MS analysis. Therefore, sample preparation protocols incorporate steps to remove or reduce denaturing reagents from the extracted proteins before trypsin digestion. The classical approach is to run the proteins on a SDS-polyacrylamide gel, and digest them in-gel. The tryptic peptides diffused from the gel are collected for LC-MS/MS analysis [1] (for detailed description, *see* Chapters 8 and 11). Although this protocol is robust, it is time-consuming and labour-intense. Furthermore, to collect peptides that passively diffused out of the gel for LC-MS/MS analysis may not be optimal for peptide recovery. Recently, several protocols that facilitate the treatment of samples in a single vessel have been developed [2–9]. These protocols minimize processing steps, improve peptide recovery, and have the potential for large-scale analysis in which multiple samples are processed in parallel in individual wells of a 96-well plate [10]. FASP is the most widely used protocol, whereas the SP3 protocol is considerably faster, cheaper, and recently proven to be compatible with a wide range of detergents for sample preparation [5]. The present chapter gives experimental details on SP3 and FASP [11] protocols for the analysis of isolated synaptosome samples.

SP3, introduced in 2014 by Prof. Krijgsveld's laboratory [4], is based on the use of beads with hydrophobic interaction functionality via, for example, surface coating with carboxylate groups. The extracted proteins are bound to the beads at high organic solvent concentration. The contaminants including denaturing reagents are effectively removed with several washing rounds with ethanol and acetonitrile. Proteins are digested on-bead by trypsin in an aqueous buffer. The absence of organic solvent concomitantly releases proteins/peptides from the beads into the digestion buffer. All the handling steps are performed in a single reaction tube, which minimizes sample loss, and the peptides in principle can be collected in a small volume (down to 10 μL) for direct loading into the LC-MS/MS. Taken together, SP3 opens up an avenue for ultrasensitivity proteomics, for example, the single cell type analysis. For high-throughput experiments, SP3 can also be performed in a 96-well plate format [12].

FASP [3] gained popularity in the past decade due to its ease to perform. It is executed in a filtration vial with MW cut-off of 30 kDa. The low molecular weight contaminants in the protein extracts, such as SDS and salts, are removed from the filtration vial by centrifugation and discarded. The retained proteins in the filtration vial are digested by trypsin. The resulting peptides are small enough to pass through the filter, and are collected by centrifugation. It has been reported that the filter could absorb small amount of proteins causing sample loss. This problem can be alleviated by the use of a sufficient amount of starting material, for example >10 μg. For ultrasensitivity proteomics experiments, the use of the SP3 protocol is preferred.

2 Materials

2.1 Apparatus

1. Shaker, for example the Thermoshaker from Eppendorf.
2. Magnetic rack. To facilitate the immobilization of beads a magnetic rack with strong magnet is preferable, for example the epimagHT magnetic separator from Epigentek and the 96S super magnet plate from Alpaqua.
3. LC-MS/MS. Any nano- or micro-HPLC coupled to a high-end mass spectrometer can be used. The sensitivity and analysis time depend on the type of system used.
4. Centrifuge. Any Table centrifuge can be used, for example, the Eppendorf centrifuge 5917R.
5. Speedvac. For example the Eppendorf concentrator plus.

2.2 Reagents

1. Water; it should be double deionized.
2. 2% SDS buffer; 2% SDS in 50 mM Tris-buffer pH 8. This can be prepared from the stock solutions used for casting the SDS-PAGE gel. Add 50 μL 1.5 M pH 8.8 Tris buffer and 950 μL 1.5 M pH 6.8 Tris buffer to a 50 mL falcon tube. Add 6 mL 10% SDS stock solution and make up the solution to 30 mL with de-ionized water.
3. 50 mM ammonium bicarbonate (0.195 g ammonium bicarbonate in 50 mL water).
4. Trypsin or Trypsin/endo-lysC mixture (MS sequence-grade from Promega, dissolve 1 vial of 20 μg proteases in 3 mL 50 mM ammonium bicarbonate).
5. 8 M urea (prepare fresh: put 19.2 g urea in a 50 mL Falcon tube, add 2 mL 1.5 M Tris buffer pH 8 and fill up to 40 mL. Rotate the tube at room temperature to dissolve all urea).
6. SP3 bead (two different Sear-Mag speed Beads are available from Thermo Scientific [1] catalog number 45152101010250 and [2] catalog number 65152105050250).
7. Centrifugal filter unit (Microcon-30 from Millipore).
8. 50 mM Tris (2-carboxyethyl)phosphine (TCEP). Dilute the 500 mM TCEP solution from sigma-aldrich to 50 mM with water, and store at −20 °C in small aliquot of 100 μL in a 500 μL Eppendorf tube.

9. 200 mM methyl methanethiosulfonate (MMTS). Add 1 mL isopropanol to 19.25 µL MMTS (97% purity from Fluke), store at −20 °C in small aliquot of 50 µL in a 500 µL Eppendorf tube.

3　Methods

3.1　Solubilization of Sample in 2% SDS Buffer

1. Prepare synaptosome as described in Chapter 3 (*see* **Note 1**).

2. Transfer 10–20 µg synaptosome to an Eppendrof tube, add 15 µL 2% SDS solution and 1 µL of 50 mM Tris (2-carboxyethyl) phosphine (*see* **Note 2**).

3. Shake the tube at 50 °C, 1000 rpm, for 30 min in a Thermoshaker.

4. Add 0.5 µL of 200 mM methyl methanethiosulfonate, shake at room temperature in the Thermoshaker for 5 min.

3.2　SP3 Protocol

3.2.1　Prepare SP3 Beads

1. Pipette 20 µL of the two different Sear-Mag speed Beads each into an Eppendorf tube. Make sure the beads are well suspended in solution by shaking the bottle before they are pipetted and transferred to the Eppendorf tube.

2. Add 160 µL water to the beads, vortex mildly.

3. Place the tube on a magnetic raft. The beads migration toward the magnetic side may take several minutes to complete.

4. Remove the solution and discard, add 200 µL water to the beads.

5. Repeat the rinsing step once.

6. Store the beads in 100 µL water at 4 °C.

3.2.2　Sample Preparation with SP3 Protocol

1. Add 5 µL SP3 bead mix to the solubilized synaptosome sample of about 15 µL (10–20 µg), and then add 30 µL 100% ethanol, vortex gently to mix solutions. When a larger volume of sample is used, add more ethanol and maintain the ethanol/water ratio >60%.

2. Incubate for 5 min at room temperature with occasional gentle shaking of the tube by hand.

3. Place the samples on the magnetic rack (*see* **Note 3**).

4. When all beads have migrated to the tube side against the magnet, pipette and discard the solvent.

5. Add 200 µL 70% ethanol and incubate for few seconds on magnetic raft, then remove solvent completely.

6. Repeat this washing step with 70% ethanol at least three times.

7. Repeat washing of the beads in 100% acetonitrile four times.

8. After discarding solvent, add 100 μL of trypsin solution (*see* **Note 4**).

9. Incubate the tube at 37 °C overnight.

10. Place the tube on the magnetic rack to immobilize the beads (*see* **Note 5**).

11. Transfer the supernatant to a vial, freeze dry and store at −80 °C.

12. Re-dissolve the dried peptides in appropriate volume of HPLC solvent A for LC-MS/MS analysis.

3.3 FASP Protocol

1. Mix the solubilized synaptosome sample with 100 μL freshly prepared 8 M urea solution, and transfer to a microcon-30 centrifugal filter unit.

2. Centrifuge at 13,500 × g for 14 min at room temperature, discard the flow-through (*see* **Note 6**).

3. Repeat this washing step with 200 μL 8 M urea four times, and discard the flow through.

4. Add 200 μL 50 mM ammonium bicarbonate to the filter, centrifuge for 10 min to remove urea solvent from the filter unit.

5. Repeat the washing step four times.

6. Add 100 μL trypsin to the filter. This corresponds to about 0.6 μg proteases per sample.

7. Replace the underlying tube with a new clean tube.

8. Place the whole unit in a humid chamber (for example, in a closed box with a piece of water-wetted tissue paper), incubate at 37 °C overnight.

9. Centrifuge at 14,000 × g for 20 min.

10. Add 50 μL 0.1 M acetic acid to the filter and centrifuge at 14000 × g for 20 min.

11. Transfer the peptide-rich eluate in the underlying tube to an eppendorf tube, and dry in a speedvac.

12. Store the tube containing the peptides at −80 °C until used for MS analysis.

3.4 LC-MS/MS Analysis

1. Sample can be analyzed by any modern LC-MS/MS system. For general description *see* Chapters 5 and 14; for data-dependent acquisition (DDA) *see* Chapters 5, 8, 9, and 14; for data-independent acquisition (DIA, also called SWATH) *see* Chapter 11.

4   Notes

1. Protein concentration can be determined by, for example, Bradford analysis. For quantitative proteomics, it is important to use equal amounts of proteins for all samples. We routinely run samples on a 10% SDS-PAGE gel, stain with Coomassie blue or activate TCE by UV light if proteins are run on a stain-free gel, and then quantify total proteins with a scanner (for example, the Gel Doc EZ system from Biorad that also can activate the TCE), and accordingly adjust the protein amounts across all samples for sample preparation.

2. Any biological sample can be processed. For the analysis of total tissue/cell lysate, DNA/RNA should be degraded by, for example, Benzonase Nuclease before proceeding with FASP or SP3. Alternatively, the gel-based sample preparation protocol could be used (*see* Chapters 4 and 11).

3. For samples to be processed in a 96-well plate, we use Microplate 96/V-PP from Eppendorf and 96S super magnet plate from Alpaqua. Alpaqua is a 96 ring-shaped magnetic plate in which Sera-Mag Speed bead is captured around the perimeter of the V-shaped bottom of the 96-well plate. This creates a bead free center for easy aspiration of solution. To cap the plate we use the Sealing Mat for DWP 96/1000 from Eppendorf.

4. For ultrasensitive analysis, sample may be digested in small volume of trypsin solution, for example in 10 μL. A higher trypsin concentration should be used to accommodate the smaller solution volume. The recovered peptides can be directly loaded into the LC-MS/MS, thereby avoiding sample loss due to the extra steps of drying followed by re-dissolving sample in a smaller volume compatible for loading into the LC-MS/MS.

5. When pipetting the aqueous solution, a small amount of SP3 beads may be pulled away from the magnet by capillary action of water and contaminate the collected peptide solution. These SP3 beads can be removed by a desalting step using a Stage-tip or Ziptip, or with a 0.2 μm filter to retain the beads. We routinely use the Pall Acroprep as filter (0.2 μm GHP membrane in a 96-well plate format, product ID 8082). The eluted peptides can then be speedvac dried.

6. It is possible that the filtration rates differ among the samples. Some samples may require longer centrifugation time. In extreme cases the filter may be blocked. Solution in the filter can be transferred to a new filter and one can continue with the remaining washing steps.

References

1. Chen N, Pandya NJ, Koopmans F, Castelo-Szekelv V, van der Schors RC, Smit AB, Li KW (2014) Interaction proteomics reveals brain region-specific AMPA receptor complexes. J Proteome Res 13:5695–5706

2. Sielaff M, Kuharev J, Bohn T, Hahlbrock J, Bopp T, Tenzer S, Distler U (2017) Evaluation of FASP, SP3, and iST protocols for proteomic sample preparation in the low microgram range. J Proteome Res 16:4060–4072

3. Wisniewski JR, Zougman A, Nagaraj N, Mann M (2009) Universal sample preparation method for proteome analysis. Nat Methods 6:359–362

4. Hughes CS, Foehr S, Garfield DA, Furlong EE, Steinmetz LM, Krijgsveld J (2014) Ultrasensitive proteome analysis using paramagnetic bead technology. Mol Syst Biol 10:757

5. Moggridge S, Sorensen PH, Morin GB, Hughes CS (2018) Extending the compatibility of the SP3 paramagnetic bead processing approach for proteomics. J Proteome Res 17:1730–1740

6. Kulak NA, Pichler G, Paron I, Nagaraj N, Mann M (2014) Minimal, encapsulated proteomic-sample processing applied to copy-number estimation in eukaryotic cells. Nat Methods 11:319–324

7. Taoka M, Fujii M, Tsuchiya M, Uekita T, Ichimura T (2017) A sensitive microbead-based organic media-assisted method for proteomics sample preparation from dilute and denaturing solutions. ACS Appl Mater Interfaces 9:42661–42667

8. Ludwig KR, Schroll MM, Hummon AB (2018) Comparison of in-solution, FASP, and S-trap based digestion methods for bottom-up proteomic studies. J Proteome Res 17(7):2480–2490

9. Zougman A, Selby PJ, Banks RE (2014) Suspension trapping (STrap) sample preparation method for bottom-up proteomics analysis. Proteomics 14:1006–1000

10. Yu Y, Bekele S, Pieper R (2017) Quick 96FASP for high throughput quantitative proteome analysis. J Proteome 166:1–7

11. Pandya NJ, Klaassen RV, van der Schors RC, Slotman JA, Houtsmuller A, Smit AB, Li KW (2016) Group 1 metabotropic glutamate receptors 1 and 5 form a protein complex in mouse hippocampus and cortex. Proteomics 16:2698–2705

12. Gonzalez-Lozano MA, Koopmans F, Paliukhovich I, Smit AB, Li KW (2019) A fast and economical sample preparation protocol for interaction proteomics analysis. Proteomics:e1900027

Chapter 7

Enrichment of Low-Molecular-Weight Phosphorylated Biomolecules Using Phos-Tag Tip

Eiji Kinoshita, Emiko Kinoshita-Kikuta, and Tohru Koike

Abstract

In this chapter, we provide a standard protocol for phosphate-affinity column chromatography for the separation of phosphorylated and nonphosphorylated biomolecules by using a phosphate-binding zinc(II) complex of 1,3-bis(pyridin-2-ylmethylamino)propan-2-olate (Phos-tag). A 200-µL micropipette tip containing 10 µL of swollen agarose beads functionalized with Phos-tag moieties (Phos-tag Tip) was prepared. All steps in the phosphate-affinity separation (binding, washing, and elution) were conducted by using aqueous buffers at neutral pH values. The entire separation protocol required less than 30 min per sample. This micropipette-tip method would be thus used preferentially as an alternative to existing tools for the reliable enrichment of phosphorylated biomolecules, such as phosphopeptides, in the field of neuroscience.

Key words Affinity chromatography, Micropipette-tip method, Phosphopeptide, Phosphoproteomics, Phos-tag

1 Introduction

Phosphorylation is one of the most important posttranslational modifications that regulate the function, localization, and binding specificity of particular proteins [1–3]. In mammalian cells, this modification occurs mainly on serine, threonine, and tyrosine residues, and is essential for the regulation of life. Changes in the phosphorylation of states of proteins fundamentally affect many cellular events and are involved in numerous diseases, such as cancers or neurodegenerative disorders [4, 5]. As a result, the phosphorylated forms of proteins have become one of the major targets for clinical proteome analysis aimed at drug discovery and the production of customized medicines. Actually, protein kinase inhibitors have played an increasingly prominent role in the treatment of cancer and other diseases [6]. Rapid and specific enrichment of native phosphoproteins from complex biological samples is therefore an important process in the fields of biology and medicine.

Ka Wan Li (ed.), *Neuroproteomics*, Neuromethods, vol. 146,
https://doi.org/10.1007/978-1-4939-9662-9_7, © Springer Science+Business Media, LLC, part of Springer Nature 2019

Conventional methods for enriching phosphorylated biomolecules include immobilized metal affinity chromatography (IMAC) [7, 8] and metal oxide affinity chromatography (MOAC) [9]. IMAC is based on the principle that trivalent metal ions carrying positive charges, such as Fe^{3+} or Ga^{3+}, interact with phosphate groups, which carry negative charges. MOAC, on the other hand, is based on the principle that titanium dioxide (TiO_2, titania) interacts with phosphate groups carrying negative charges under acidic conditions. However, such methods involving metal ions require the use of acidic or basic conditions, under which phosphorylated proteins can become denatured or inactivated. Therefore, although these methods can be applied in the analysis of phosphorylated peptide fragments, they cannot be used in the analysis of "natural" phosphorylated full-length proteins. In other words, IMAC and MOAC can provide information about the primary structures of proteins, but not about their tertiary structures, functions, characteristics, or behavior. Furthermore, these methods have poor specificity because the metal ions can also bind strongly to sites other than phosphate groups, such as carboxy groups.

We have been developing several techniques for phosphoproteomic studies by using a series of functional Phos-tag molecules that bind specifically to phosphate groups. Phos-tag is a binuclear metal complex that can selectively bind to a phosphate monoester in an aqueous solution at neutral pH values [10, 11]. The zinc(II) complex, which has a vacancy on the two zinc(II) ions, can interact with a phosphate monoester dianion as a bridging ligand to form a stable 1:1 complex in aqueous solution. Among the techniques, we used a Phos-tag derivative, Phos-tag agarose, to demonstrate the use of phosphate-affinity chromatographic separation, enrichment, and purification of relatively high-molecular-mass phosphorylated biomolecules such as phosphoproteins [12–14].

In this chapter, we describe a protocol using a 200-μL micropipette-tip (called the "Phos-tag Tip") containing 10 μL of swollen Phos-tag agarose beads, as a novel phosphate-affinity tool for the simple and convenient enrichment of the low-molecular-weight phosphorylated biomolecules, such as nucleotides, phosphorylated amino acids, or phosphopeptides [15].

2 Materials (*See* **Note 1**)

2.1 Solutions for Phosphate-Affinity Enrichment

1. Preservation solution (Sol. A): 20% v/v aqueous 2-propanol.

2. Binding/washing buffer (Sol. B): 0.10 M Bis-Tris/CH_3COOH buffer, pH 6.8 (*see* **Note 2**) containing 0.10 M CH_3COONa.

3. Elution buffer (Sol. C): 0.10 M $Na_4P_2O_7$/0.10 M CH_3COOH, pH 7.0 (*see* **Note 3**).

2.2 Apparatuses for Phosphate-Affinity Enrichment

1. Phos-tag Tip (nAG1-103): Commercially available from NARD Institute, Amagasaki, Japan.

2. 1.0-mL disposable plastic syringe

3. Silicone-rubber tube adaptor (8 mm long and 4 mm in diameter).

4. 1.5-mL microcentrifuge tube.

3 Methods

3.1 Preparation of Phos-Tag Tip (See Fig. 1)

1. Wash off Sol. A on both the outside and the inside of the Phos-tag Tip with 1 mL of distilled water.

2. After addition of 100 μL of Sol. B from the top of the Phos-tag Tip, attach a 1.0-mL disposable plastic syringe with a silicone-rubber tube adaptor to the Phos-tag Tip.

3. Displace the liquid in the Phos-tag Tip by air from the syringe.

4. Repeat the washing steps of Subheading 3.1, **steps 2** and **3** twice (*see* **Note 4**).

3.2 Flow-Through Fraction (FT, See Fig. 2)

1. Prepare a sample solution (30–100 μL in a 1.5-mL microcentrifuge tube) containing ≤50 nmol phosphorylated species in Sol. B. Keep the sample solution with the range of pH values from 6 to 8.

2. Draw the sample solution gently into the Phos-tag Tip by using the syringe and keep the Phos-tag agarose beads in suspension for a few seconds.

3. Pass all the sample solution through the Phos-tag agarose beads and then move the liquid above the upper filter.

4. Displace gently the liquid in the Phos-tag Tip by air from the syringe into the microcentrifuge tube used for the sample preparation.

5. Repeat the phosphate-binding steps from Subheading 3.2, **steps 2–4** five times.

6. Collect the liquid as the flow-through (FT) fraction in the microcentrifuge tube.

3.3 Washing Fractions (W1, W2, W3, and W4, See Fig. 3)

1. After removing the syringe, add 100 μL of Sol. B to the Phos-tag Tip from the top.

2. Attach the syringe again and push gently Sol. B into the space between the two filters.

3. Resuspend the phosphate-bound Phos-tag agarose beads in the Phos-tag Tip by gently moving the syringe piston up and down a few times.

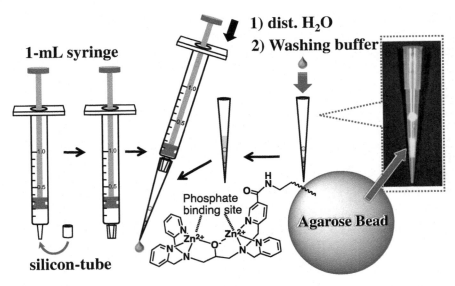

Fig. 1 Preparation of Phos-tag tip. Inset shows a photograph of Phos-tag Tip before the operation of chromatography

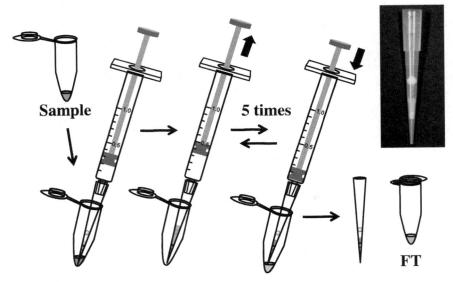

Fig. 2 Binding of phosphate-containing molecules. Inset shows a photograph of Phos-tag tip, in which Phos-tag agarose beads capture riboflavin 5′-phosphate used as a sample. For experimental details *see* ref. 15

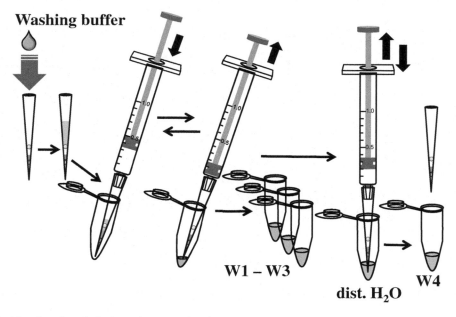

Fig. 3 Washing the phosphate-bound agarose beads

4. Displace the liquid in the Phos-tag Tip by air from the syringe into a microcentrifuge tube.

5. Repeat the washing steps of Subheading 3.3, **steps 1** and **5** three times to obtain three washing fractions (W1, W2, and W3).

6. Prepare a microcentrifuge tube containing 100 µL of distilled water.

7. Draw the water gently into the Phos-tag Tip and move the liquid above the upper filter.

8. Displace all the liquid in the Phos-tag Tip by air from the syringe into the microcentrifuge tube to obtain a washing fraction (W4).

3.4 Elution Fractions (E1, E2, and E3, See Fig. 4)

1. After removing the syringe, add 50–100 µL of Sol. C to the Phos-tag Tip from the top.

2. Attach the syringe again and push gently Sol. C into the space between the two filters.

3. Resuspend the phosphate-bound Phos-tag agarose beads in the Phos-tag Tip by gently moving the syringe piston up and down a few times.

4. Displace all the liquid in the Phos-tag Tip by air from the syringe into a microcentrifuge tube to obtain the first elution fraction (E1).

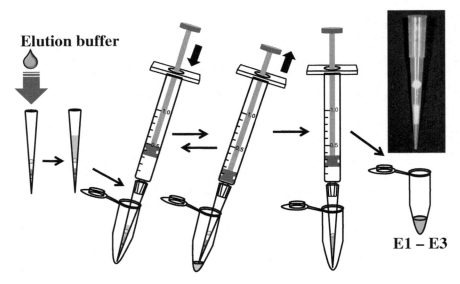

Elution buffer

E1 – E3

Fig. 4 Eluting the phosphate-containing molecules. Inset shows a photograph of Phos-tag Tip, in which Phos-tag agarose beads release riboflavin 5′-phosphate (*see* Fig. 2). For experimental details *see* ref. 15

5. If necessary, repeat the elution steps of Subheading 3.4, **steps 1** and **4** to obtain additional elution fractions (E2 and E3) (*see* **Note 5**).

3.5 Typical Results for the Separation of Phosphopeptides from Cell Extracts

By using Phos-tag Tip containing Phos-tag agarose beads (1 nmol/μL of phosphate-binding site, 10 μL), followed by LC-MS/MS analyses, we identified 1649 non-redundant phosphopeptides from the lysates of HEK293 cells (the peptides sample derived from 25 μg proteins per MS analysis) in three independent experiments (Fig. 5a) [15]. The average ratio of identified phosphopeptides to total peptides in each experiment was over 90%, indicating that the specificity is high. The high reproducibility was also demonstrated in the triplicate enrichments. A large overlap in the identified phosphopeptides was obtained by simple repetitions using Phos-tag tip, as can be seen in the Venn diagrams. The triplicate analyses permitted us to identify a total of 3198 phosphopeptides, containing 668 unique phosphopeptides and 2530 identical phosphopeptides, indicating that ~80% of identical phosphopeptides can be detected reproducibly. In addition, the high correlation between the analyses was confirmed by scatter plots based on the normalized abundance of each peptide, as calculated by label-free peptide relative quantification analysis in Progenesis QI (Fig. 5b). Sample preparation for MS-based analysis is a critical step in quantitative proteomics. Therefore, the high reproducibility of the selective phosphopeptide enrichment suggests that the use of a Phos-tag tip

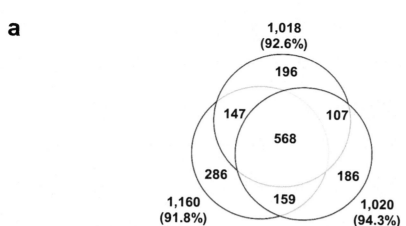

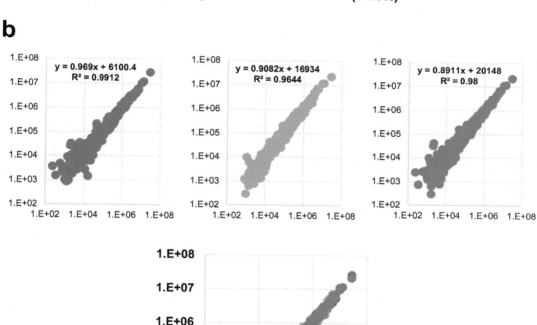

Fig. 5 Enrichment of phosphopeptides by using Phos-tag Tip. (**a**) The overlaps of identified phosphopeptides among three different samples, obtained by simple repetitions using Phos-tag Tip (1 nmol/μL of phosphate-binding site) followed by LC-MS/MS analysis. The numbers indicate identified peptides and the percentages indicate the enrichment efficiency of phosphopeptides [(number of phosphopeptides/number of identified peptides) × 100]. The peptides sample derived from 25 μg proteins was subjected to one LC-MS/MS analysis. Fig. 5 (continued) (**b**) Correlation of the normalized abundances of each peptide calculated by label-free peptide relative quantification analysis. The three upper panels show the correlations between the two experiments, and the lower panel shows an overlay of the three upper panels. For experimental details *see* ref. 15. Reprinted with permission from *Electrophoresis* [15] Copyright (2017) Wiley-VCH Verlag GmbH & Co. KGaA, Weinheim

Phos-tag Tip

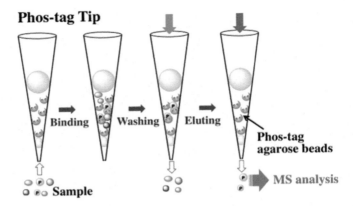

Binding   Washing   Eluting

Phos-tag agarose beads

MS analysis

Sample

Fig. 6 Summary of phosphate-affinity separation using Phos-tag Tip

might become a standard technique in MS-based quantitative phosphorylation proteomics in the field of neuroscience.

We described a protocol for simple and efficient separation of phosphopeptides by using a phosphate-affinity micropipette-tip method. The phosphate-affinity site in a micropipette tip is an alkoxide-bridged dinuclear zinc(II) complex of 1,3-bis(pyridin-2-ylmethylamino)propan-2-olate (Phos-tag), which is linked to hydrophilic crosslinked agarose. The Phos-tag tip contains 10 μL of swollen Phos-tag agarose beads, and it is used in conjunction with a 1.0-mL syringe attached with an adapter made of silicone tubing. This method is categorized as an IMAC separation. The exchange reaction of phosphate ligands on the Zn^{2+} ions in the Phos-tag moieties is fast; consequently, the binding, washing, and elution steps each require a short time of less than 2 min. All the binding, washing, and elution steps in the phosphate-affinity separation process are conducted in aqueous buffers at neutral pH values (summarized in Fig. 6). As well as requiring short operation times, the method reduces damage to the separated molecules. We expect that this Phos-tag tip technique would be used preferentially as an alternative to existing tools for the reliable enrichment of phosphopeptides in phosphoproteome study in the field of neuroscience.

4   Notes

1. All reagents and solvents used should be purchased at the highest commercial quality available and used without further purification. All aqueous solutions should be prepared using deionized and distilled water.

2. Acetic acid (CH_3COOH) is dangerously irritating to the skin, eyes, and mucous membranes. When handing this chemical, work in a chemical fume hood and wear gloves, eye protection, and a mask. Keep away from heat and flame.

3. Other elution buffers (e.g., 0.1 M aqueous HCl containing 2% v/v aqueous TFA and 5% v/v aqueous NH_3 or an aqueous solution containing 0.10 M EDTA/NaOH, pH 7.0) can be used.

4. Check that Phos-tag agarose beads (10 μL) are compressed on the lower filter in the Phos-tag Tip.

5. To reduce the amount of liquid remaining in the micropipette tip, we recommend the use of a microcentrifuge (e.g., $2000 \times g$ for 20 s) and a 1.5-mL microcentrifuge tube attached to an appropriate tip adaptor. For mass spectrometric analysis, the phosphate-enriched sample can be desalted by a commonly used method involving a reverse-phase resin.

Acknowledgments

This work was supported in part by KAKENHI Grants 25293005 to E.K., 15 K07887 to E.K.-K., and 26460036 to T.K.

References

1. Hunter T (2000) Signaling: 2000 and beyond. Cell 100:113–127

2. Olsen JV, Blagoev B, Gnad F, Macek B, Kumar C, Mortensen P, Mann M (2006) Global, in vivo, and site-specific phosphorylation dynamics in signaling networks. Cell 127:635–648

3. Ubersax JA, Ferrell JE Jr (2007) Mechanisms of specificity in protein phosphorylation. Nat Rev Mol Cell Biol 8:530–541

4. Brognard J, Hunter T (2011) Protein kinase signaling networks in cancer. Curr Opin Genet Dev 21:4–11

5. Martin L, Latypova X, Terro F (2011) Post-translational modifications of tau protein: implications for Alzheimer's disease. Neurochem Int 58:458–471

6. Gross S, Rahal R, Stransky N, Lengauer C, Hoeflich KP (2015) Targeting cancer with kinase inhibitors. J Clin Invest 125:1780–1789

7. Andersson L, Porath J (1986) Isolation of phosphoproteins by immobilized metal (Fe^{3+}) affinity chromatography. Anal Biochem 154:250–254

8. Posewitz MC, Tempst P (1999) Immobilized gallium(III) affinity chromatography of phosphopeptides. Anal Chem 71:2883–2892

9. Sano A, Nakamura H (2004) Titania as a chemo-affinity support for the column-switching HPLC analysis of phosphopeptides: application to the characterization of phosphorylation sites in proteins by combination with protease digestion and electrospray ionization mass spectrometry. Anal Sci 20:861–864

10. Kinoshita E, Takahashi M, Takeda H, Shiro M, Koike T (2004) Recognition of phosphate monoester dianion by an alkoxide-bridged dinuclear zinc(II) complex. Dalton Trans 1189–1193

11. Kinoshita E, Kinoshita-Kikuta E, Takiyama K, Koike T (2006) Phosphate-binding tag, a new tool to visualize phosphorylated proteins. Mol Cell Proteomics 5:749–757

12. Kinoshita E, Yamada A, Takeda H, Kinoshita-Kikuta E, Koike T (2005) Novel immobilized zinc(II) affinity chromatography for phosphopeptides and phosphorylated proteins. J Sep Sci 28:155–162

13. Kinoshita-Kikuta E, Kinoshita E, Yamada A, Endo M, Koike T (2006) Enrichment of phosphorylated proteins from cell lysate using a novel phosphate-affinity chromatography at physiological pH. Proteomics 6:5088–5095

14. Kinoshita-Kikuta E, Kinoshita E, Koike T (2009) Phos-tag beads as an immunoblotting enhancer for selective detection of phosphoproteins in cell lysates. Anal Biochem 389:83–85

15. Yuan ET, Ino Y, Kawaguchi M, Kimura Y, Hirano H, Kinoshita-Kikuta E, Kinoshita E, Koike T (2017) A Phos-tag-based micropipette-tip method for rapid and selective enrichment of phosphopeptides. Electrophoresis 38:2447–2455

Chapter 8

Integrated Immunoprecipitation: Blue Native Gel Electrophoresis—Mass Spectrometry for the Identification of Protein Subcomplexes

Sophie J. F. van der Spek, August B. Smit, and Nikhil J. Pandya

Abstract

Proteins typically act as components of poly-protein complexes, which may vary in composition and function, and may depend on subcellular localization and cellular state. To understand their biology, there is a need to identify the constituents in these various protein complexes. Conventional immunoprecipitation combined with proteomics analysis is nowadays routinely used to identify multiple interactors of a bait protein; however, it is unable to distinguish protein subcomplexes from one another. Blue native polyacrylamide gel electrophoresis is appropriate to separate individual protein complexes. Here, we describe a protocol that combines the specificity of immunoprecipitation, the size separating capacity of Blue native polyacrylamide gel electrophoresis, and subsequent mass spectrometry to delineate the protein constituents in subcomplexes of a targeted protein.

Key words Immunoprecipitation, Blue native polyacrylamide gel electrophoresis, Protein complexes, Mass spectrometry, Proteomics

1 Introduction

Proteins typically occur as components of poly-protein complexes. These may exist in a vast number of different forms each with distinct protein composition and abundances, driving biological processes with specific spatial and temporal patterns. Their accurate mapping may provide important insights into the functioning of the cell. To identify stably interacting proteins, immunoprecipitation combined with mass spectrometry (IP-MS) (Fig. 1a) is the method of choice, in which knock-out or knock-down negative controls are often included to rule out false positives [1–3].

IP-MS has been instrumental in determining stable protein complexes from the brain, for example over 30 interacting proteins have been described for the AMPA receptor [4, 5]. Protein assemblies, such the AMPA-receptor and its variety of interactors, exist in multiple subcomplexes, which may have different combinations

Ka Wan Li (ed.), *Neuroproteomics*, Neuromethods, vol. 146,
https://doi.org/10.1007/978-1-4939-9662-9_8, © Springer Science+Business Media, LLC, part of Springer Nature 2019

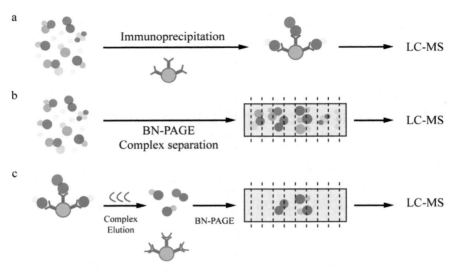

Fig. 1 Strategies for protein complex identification. (**a**) Conventional immunoprecipitation strategy involving extracted (left) immunoprecipitation of protein subcomplexes for the bait protein (in blue). Pull down of the bait protein is unable to distinguish between protein subcomplexes as identified with LC-MS. (**b**) Blue native page separation of protein extracts followed by gel slicing and LC-MS to identify separated protein complexes. (**c**) Combined IP-BN-MS strategy involving elution of protein complexes post IP using peptide antigen, followed by BN-PAGE separation of purified complexes and LC-MS detection and identification of protein subcomplexes

of constituents throughout the life-time of the receptor, different cell-types, subcellular compartments, and cellular states [6]. Although conventional IP-MS is a powerful tool to identify multiple (novel) interactors of a bait protein at once, and has been instrumental for delineating the AMPAR interactome, this technique alone does not allow distinguishing protein subcomplexes at an individual basis.

Methods for separation of intact protein complexes include size exclusion chromatography, ion exchange chromatography or gel-based approaches, such as blue native polyacrylamide gel electrophoresis (BN-PAGE) [7, 8]. In particular BN-PAGE has been demonstrated to separate intact protein complexes with adequate resolution [9], and in combination with subsequent mass spectrometry can be used to identify the constituents of individual complexes (Fig. 1b, [10]).

There are two critical technical issues with BN-PAGE. Highly complex samples, such as a brain extract, may contain a high number of protein complexes many of which have similar masses and exhibit similar migration patterns on BN-PAGE. Co-migration in BN-PAGE may limit the unequivocal identification of proteins belonging to a distinct subcomplex. Second, a single protein may be present in multiple protein complexes of different composition. These protein subcomplexes may spread out or migrate to multiple spots on the gel probably with different intensities (Fig. 1b [10]), making the identification of the subcomplex to which these belong challenging. Characterization of protein subcomplexes therefore typically requires a fractionation step in addition to BN-PAGE that preferably removes the bulk of non-related complexes and thus enriches the protein complex of interest to facilitate the downstream analysis.

A solution to the problem of identification of distinct protein complexes amid high sample complexity and wide MW range of protein complexes using BN-PAGE is to combine the specificity of IP experiments and the separation offered by BN-PAGE into one single workflow, called IP-BN-MS (Fig. 1c, *see* also [11]). In this workflow we enrich for protein complexes of interest by performing a classical IP against the bait protein using an antibody for which the antigenic sequence is available. Subsequently, we elute the bait and its subcomplexes in their native state using a high concentration of the peptide containing the known antibody epitope. The complexity of protein complexes loaded on the gel is thus greatly reduced by this IP pre-isolation step. The complexes left are then separated by size on a BN-PAGE gel, which is cut into multiple gel slices that are then prepared and analyzed for mass spectrometry individually. Proteins that are together in the same (sub)complex will co-migrate at the same location on the gel, allowing to determine its protein composition.

In this chapter, we describe an optimized procedure for performing IPs on brain homogenate followed by peptide-elution of protein complexes, separation on BN-PAGE, and their further downstream identification and quantification using proteomics.

2 Materials

2.1 Apparatus

1. Tissue homogenizer.
   Homgenplus Semi-Automatic Homogenizer (Schuett Biotech).

2. Tube rotator.
   RM-2 L Intelli-mixer (Elmi).

3. 1.5 mL tube centrifuge.
   Centrifuge 5417R with a F45-30-11 rotor (Eppendorf).

4. 0.5 mL centrifugal filters.
   Vivaspin 500, VS0121, MWCO 30 kDa (Sartorius stedim biotech).

5. 15 mL tube/96well plate centrifuge.
   Centrifuge 5810R with a A-4-62 swing bucket rotor, with rectangular and MPT buckets (Eppendorf).

6. 96-well filter plate.
   EMD Millipore™ MultiScreen HTS Durapore™ 96-Well Filter Plates (Millipore).

7. Blue Native Gel System.
   NativePAGE Novex Bis-Tris Gel System (Invitrogen).

8. Blue Native precast gels.
   NativePAGE 3-12% Bis-Tris Protein Gels, 1.0 mm, 10 well (Invitrogen).

9. Grid cutter.
   GridCutter Blades (1.5 mm × 5.0 mm) MEE1.5-5-48 and Mount MEF72-5 for 48 gel-pieces, or
   GridCutter Blades (1.0 mm × 5.0 mm) MEE1-5-50 and Mount MEF50-5, for a total of ~70 gel-pieces (Gel company).

10. SpeedVac.
    Concentrator plus with a F45-70-11 rotor (Eppendorf).

11. HPLC system.
    Ultimate 3000 LC system (Dionex, Thermo Scientific).

12. High-resolution mass spectrometer.
    TripleTOF 5600 (Sciex).

2.2 Software

1. Maxquant.

2.3 Reagents

All buffers prepared in MilliQ quality water.

2.3.1 Immuno-Purification

1. Extraction solution:
   25 mM HEPES/NaOH, pH 7.4.
   150 mM NaCl.
   1% *n*-Dodecyl-beta-Maltoside (DDM).
   EDTA-free protease inhibitor cocktail (Roche).

2. 100 µg antibody.

3. Protein A/G PLUS-Agarose (Santa Cruz).

4. Wash solution:
   25 mM HEPES/NaOH, pH 7.4.
   50 mM NaCl.
   0.1% Triton-X.

5. Peptide solution:
   0.5 mg/mL peptide corresponding to the antibody epitope dissolved in wash solution. For each incubation step with peptide solution, the peptide should be ~250× more abundant in moles than the antibody. The concentration provided here is calculated based on a peptide with a length of 20 amino acids.

6. BN-PAGE reagents (Invitrogen).
   NativePAGE™ 5% G-250 Sample Additive.
   NativePAGE™ Sample Buffer (4×).
   NativePAGE™ Cathode Buffer Additive (20×).
   NativePAGE™ Running Buffer (20×).

Use the manual instructions for preparing the cathode and anode buffers.

7. Fixation solution:
   50% ethanol.
   3% phosphoric acid.

8. Colloidal Coomassie:
   34% methanol.
   3% phosphoric acid.
   15% ammonium sulfate.
   0.1% Coomassie brilliant blue G-250.

2.3.2 In Plate Digestion

1. 50 mM ammonium bicarbonate.

2. 50 mM ammonium bicarbonate, 0.5 mM Tris(2-carboxyethyl) phosphine hydrochloride (TCEP).

3. 50 mM ammonium bicarbonate, 1 mM methyl methanethiosulfonate (MMTS).

4. 50 mM ammonium bicarbonate/50% acetonitrile.

5. 100% acetonitrile.

6. Trypsin solution: Sequencing Grade Modified Trypsin (Promega). Add 3 mL 50 mM ammonium bicarbonate to the vial containing trypsin, and vortex briefly.

7. 0.1% Trifluoroacetic acid/50% acetonitrile.

8. 0.1% Trifluoroacetic acid/80% acetonitrile.

3 Methods

3.1 Immuno-Purification

To ensure sample integrity all steps should be performed at 4 °C.

3.1.1 Protein Extraction and Antibody Incubation

1. Homogenize frozen brains in the 1% DDM extraction solution using a homogenizer set at 900 rpm for 12 strokes at 4 °C.

   • Use 5 cortices or 20 hippocampi (both hemispheres) in 15 mL (*see* **Note 1**).

2. Incubate on a rotator at 10 rpm for 1 h at 4 °C.

3. Centrifuge twice at $20,000 \times g$ for 20 min at 4 °C.

   • Use 2 mL tubes for centrifugation.

   • Collect final supernatants in a 15 mL tube.

4. Incubate the supernatant with 100 μg of antibody on a rotator at 10 rpm overnight at 4 °C (*see* **Note 2**).

3.1.2 Pull Down

1. Wash 1 mL of protein A/G beads in 10 mL wash solution in a 15 mL tube.

2. Centrifuge at $1000 \times g$, 2 min at 4 °C and discard the supernatant.

3. Repeat **steps 1** and **2** once.

   • Make sure you add the wash solution on top of the beads to displace them.

4. Add the sample to the beads.

5. Incubate on a rotator at 10 rpm for 1 h at 4 °C.

6. Centrifuge at $1000 \times g$, 2 min at 4 °C and discard the supernatant.

7. Add 10 mL wash solution.

   • Try to carefully displace the beads while adding the wash solution.

8. Centrifuge $1000 \times g$, 2 min at 4 °C and discard supernatant.

9. Repeat **steps 7** and **8** three times.

   • While removing supernatant the last time keep the total volume at 1.5 mL.

10. Mix careful with a pipette and transfer the beads to a 1.5 mL tube.

11. Centrifuge $1000 \times g$, 1 min at 4 °C and discard supernatant.

3.2 Peptide Elution and Sample Volume Reduction

1. Precool the BN-PAGE cathode and anode buffer at 4 °C.

2. Add 1 mL of peptide solution to the Protein A/G beads.

3. Incubate on rotator at 10 rpm, 45 min at 4 °C.

4. Centrifuge $1000 \times g$, 1 min at 4 °C.

   • Collect eluate in a 2 mL tube.

5. Repeat **steps 2–4**.

6. Centrifuge the sample at $10,000 \times g$, 1 min at 4 °C.

   • To remove accidently pipetted beads.

7. Load 500 μL of the sample on a Vivaspin 500 filter tube.

8. Centrifuge at $14,000 \times g$, 10 min at 4 °C and discard the flow through.

9. Repeat **steps 6** and **7** until the volume is ~30 μL or less.

3.3 Running of BN-PAGE and Fixation of the Gel

1. Transfer the sample retained on the Vivaspin 500 filter to a 1.5 mL tube.

   • Use a gel-loading tip.

2. Mix 10 μL 4× BN loading buffer with 1 μL 5% G-250 Coomassie and 0.5 μL molecular weight marker (leave out for western blotting).

- Beforehand centrifuge 5% G-250 Coomassie at $10,000 \times g$ for 2 min to avoid loading particles.

3. Add the loading mix to the sample and mix with a pipette.

4. Centrifuge at $20,000 \times g$, 2 min at 4 °C.

5. Prepare the BN-PAGE gel system.

- Check for leakage by adding a layer of cathode running buffer in the gel compartment between the gel and the plastic buffer dam.

- Add cathode buffer to the wells to see the wells clearly.

6. Load the entire sample on the gel.

- Load a marker, and load empty wells with 1× BN loading buffer.

- Always leave at east one lane between two samples.

7. Fill the tank between glass plates with cathode running buffer and the outer compartment with anode runner buffer.

8. Run the gel at 150 V for 1.5 h followed by 250 V for 1 h at 4 °C or at 1 mA for 1 h followed by 2 mA overnight to a total running time of 17 h.

- Short run: gel contains proteins between 20-66 kDa and higher.

- Long run: gel contains proteins of 242 kDa and higher.

9. Take the gel out of the running system.

10. Incubate gel with fixation solution overnight (short run), or the next day for 1 h (long-run).

3.4 In-Plate Digestion Protocol for IP/BN-PAGE

All steps are performed at room temperature unless stated otherwise.

3.4.1 Cutting and Distaining of the Gel

1. Wash the gel three times with MilliQ water for 15 min.

- The gel should re-swell to its normal size.

2. Incubate the gel with Coomassie for 5 min.

- Only if necessary.

3. Wash the gel three times with MilliQ water for 15 min.

4. Use the grid cutter to cut the sample lane in 48 or 70 pieces.

5. Transfer the gel pieces to a 96-well filter plate.

6. Add 100 μL of 50 mM ammonium bicarbonate 0.5 mM TCEP solution.

7. Incubate for 30 min at 37 °C.

8. Centrifuge at $200 \times g$, 1 min.

9. Add 100 μL of 50 mM ammonium bicarbonate with 1 mM MMTS solution.

10. Incubate for 15 min.

11. Centrifuge at 200 × g, 1 min.

12. Wash with 100 μL 50 mM ammonium bicarbonate.

13. Centrifuge at 200 × g, 1 min.

14. Add 150 μL of 50 mM ammonium bicarbonate/50% acetonitrile.

15. Incubate for 20 min.

16. Centrifuge at 200 × g, 1 min.

17. Add 150 μL of 100% acetonitrile and incubate for 5 min.

   • The gel fragments should turn white and shrink.

18. Centrifuge at 200 × g, 1 min.

19. Add 100 μL 50 mM ammonium bicarbonate and incubate for 10 min.

20. Add 100 μL 50 mM ammonium bicarbonate/50% acetonitrile.

21. Incubate 20 min or overnight until the gel is completely distained.

22. Centrifuge at 200 × g, 1 min.

3.4.2 Digestion

1. Add 150 μL 100% acetonitrile and incubate for 15 min.

2. Centrifuge at 200 × g, 1 min.

3. Put a new deep bottom plate under the 96-well filter plate.

4. Re-swell the gel fragments in trypsin-buffer with 80 μL trypsin solution.

5. Add 50 μL 50 mM ammonium bicarbonate.

6. Digest overnight at 37 °C.

   • Put the 96-well plate in a closed box with a tissue wetted with water to prevent drying of the samples during incubation.

3.4.3 Peptide Extraction and Drying

1. Centrifuge the 96-well filter plate at 200 × g, 1 min.

2. Add 150 μL 0.1% trifluoroacetic acid/50% acetonitrile and incubate for 20 min.

3. Centrifuge at 200 × g, 1 min.

4. Add 150 μL 0.1% trifluoroacetic acid/80% acetonitrile and incubate for 20 min.

5. Centrifuge at 200 × g, 4 min.

6. Transfer the solution to 1.5 mL tubes.

7. Dry the peptide solution in a SpeedVac and store the dried peptides at −20 °C until further use.

3.5 Mass Spec Analysis

3.5.1 Dissolving Peptides for HPLC

1. Re-dissolve the sample in mobile phase A (2% acetonitrile, 0.1% formic acid).

2. Vortex.

3. Centrifuge at $10,000 \times g$, 2 min.

4. Transfer sample to a sample vial.

3.5.2 Running of Samples on a Triple TOF 5600+

1. Trap the peptides on a 5 mm Pepmap 100 C18 column (300 μm id, 5 μm particle size, Dionex).

2. Fractionate the peptides in a 200 mm Alltima C18 homemade column (100 μm ID, 3 μm particle size), using 0.1% formic acid with a linear gradient of increasing acetonitrile concentration from 5% to 30% in 35 min, to 40% at 37 min, and to 90% for 10 min at a flow rate of 500 nL/min.

3. Back equilibrate to 5% acetonitrile for 20 min.

4. Electrospray the peptides into the mass spectrometer using an ion spray voltage of 2.5 kV, ion source gas at 2 p.s.i., curtain gas at 35 p.s.i., and an interface heater temperature of 150 °C.

5. Run the MS survey scan with a range of m/z 350–1250 acquired for 250 ms.

6. Select the top 20 precursor ions for 90 ms per MS/MS acquisition, with a threshold of 90 counts.

7. Use a dynamic exclusion of 10 s, and an activated rolling CID function with an energy spread of 5 eV.

3.5.3 Data Analysis

1. Run the raw files in the latest Maxquant version against the most updated FASTA files corresponding to the species used.

   - Run all gel-fractions labeled as an individual experiment.

   - Use MMTS(c) as fixed modification.

   - Trypsin + LysC as digestion mode.

   - Activate the match-between-runs and iBAQ options.

2. Make a line graph in excel (x: gel-slices; y: iBAQ intensity value) using the proteinGroup.txt output file for the bait protein and known interactors.

3. If protein profiles look very jagged, you may opt for manual curation of protein peaks in skyline.

4 Notes

1. The immunoprecipitation section of this protocol is a scale-up version of a typical IP experiment. To give an example: for a regular IP-based interaction proteomics, use one hippocampus per experiment. The tissue can be homogenized in 1 mL extraction buffer, the bait proteins captured with 5–10 μg antibody and pulled-down with 60 μL Protein A/G beads. Proteins bound to the Protein A/G beads can be eluted by 20 μL SDS sample buffer, run on the SDS PAGE gel, trypsin-digested and subjected to LC-MS/MS analysis.

2. The antibody and the beads can be added together to the extract. The incubation time can be shortened to 2 h with similar result as overnight incubation.

References

1. Li KW, Chen N, Klemmer P, Koopmans F, Karupothula R, Smit AB (2012) Identifying true protein complex constituents in interaction proteomics: the example of the DMXL2 protein complex. Proteomics 12:2428–2432

2. Morris JH, Knudsen GM, Verschueren E, Johnson JR, Cimermancic P, Greninger AL, Pico AR (2014) Affinity purification-mass spectrometry and network analysis to understand protein-protein interactions. Nat Protoc 9:2539–2554

3. Pandya NJ, Klaassen RV, van der Schors RC, Slotman JA, Houtsmuller A, Smit AB, Li KW (2016) Group 1 metabotropic glutamate receptors 1 and 5 form a protein complex in mouse hippocampus and cortex. Proteomics 16:2698–2705

4. Chen N, Pandya NJ, Koopmans F, Castelo-Szekelv V, van der Schors RC, Smit AB, Li KW (2014) Interaction proteomics reveals brain region-specific AMPA receptor complexes. J Proteome Res 13:5695–5706

5. Schwenk J, Harmel N, Brechet A, Zolles G, Berkefeld H, Muller CS, Bildl W, Baehrens D, Huber B, Kulik A, Klocker N, Schulte U, Fakler B (2012) High-resolution proteomics unravel architecture and molecular diversity of native AMPA receptor complexes. Neuron 74:621–633

6. Brechet A, Buchert R, Schwenk J, Boudkkazi S, Zolles G, Siquier-Pernet K, Schaber I, Bildl W, Saadi A, Bole-Feysot C, Nitschke P, Reis A, Sticht H, Al-Sanna'a N, Rolfs A, Kulik A, Schulte U, Colleaux L, Abou Jamra R, Fakler B (2017) AMPA-receptor specific biogenesis complexes control synaptic transmission and intellectual ability. Nat Commun 8:15910

7. Heusel M, Bludau I, Rosenberger G, Hafen R, Frank M, Banaei-Esfahani A, Collins B, Gstaiger M, Aebersold R (2018) Complex-centric proteome profiling by SEC-SWATH-MS. bioRxiv

8. Kirkwood KJ, Ahmad Y, Larance M, Lamond AI (2013) Characterization of native protein complexes and protein isoform variation using size-fractionation-based quantitative proteomics. Mol Cell Proteomics 12:3851–3873

9. Schagger H, von Jagow G (1991) Blue native electrophoresis for isolation of membrane protein complexes in enzymatically active form. Anal Biochem 199:223–231

10. Munawar N, Olivero G, Jerman E, Doyle B, Streubel G, Wynne K, Bracken A, Cagney G (2015) Native gel analysis of macromolecular protein complexes in cultured mammalian cells. Proteomics 15:3603–3612

11. Pardo M, Bode D, Yu L, Choudhary JS (2017) Resolving affinity purified protein complexes by blue native PAGE and protein correlation profiling. J Vis Exp

Chapter 9

Analysis of Synaptic Protein–Protein Interaction by Cross-linking Mass Spectrometry

Fan Liu

Abstract

Protein–protein interactions and their dynamics define organelles' structures and drive all aspects of cellular functions. In this chapter I describe a cross-linking mass spectrometry protocol that makes use of a newly developed MS-cleavable cross-linker to identify the protein pairs based on their proximity as defined by the length of the cross-linker. Unlike other approaches, such as immunoprecipitation-based interaction proteomics or yeast-two hybrid analysis, this protocol could characterize protein–protein interactions in a more physiological setting and allows the detection of native protein complexes contained in the mouse synaptosome.

Key words Synapse, DSSO, Cross-linking, Mass Spectrometry, Protein, Protein interaction

1 Introduction

Synaptic neurotransmission is a fundamental process of the nervous system that underlies rapid communication among neurons. It is initiated when an action potential reaches the presynapse, triggers an influx of calcium ions, and causes the fusion of the docked synaptic vesicles (SV) to the membrane of the active zone. The released transmitters bind to receptors in the postsynapse and drives the postsynaptic signalling, whereas SV proteins are retrieved to form new vesicles for the next round of release [1].

Recent proteomics analysis of biochemically enriched synaptic fractions implicates the presence of more than 2000 synaptic proteins [2, 3]. However, it is not completely clear how these proteins are structurally and functionally integrated in the synapse. In their physiological context, proteins are organized into stable or dynamic molecular assemblies, forming the basis of working machineries to execute specific cellular processes with extremely high precision. One of the well-established approaches to study the connectivity of proteins and complexes is immunoprecipitation (IP)-based interaction proteomics [4, 5]. Notwithstanding its indispensable value,

Ka Wan Li (ed.), *Neuroproteomics*, Neuromethods, vol. 146,
https://doi.org/10.1007/978-1-4939-9662-9_9, © Springer Science+Business Media, LLC, part of Springer Nature 2019

several drawbacks may present in this classic approach. First, immunoprecipitation-based approaches critically depend on the availability of high-quality antibodies that are able to purify/enrich the synaptic proteins of interest with low cross-reactivity. For many synaptic proteins, especially the less-studied ones, high-quality antibodies are not immediately available from commercial sources. Second, the experimental conditions of individual IP experiment need to be carefully examined, such as buffer compositions and incubation time, which can be time-consuming and labour-intensive [4]. Third, IP experiments often require the use of detergents to solubilize proteins, especially membrane proteins from tissue preparations, such as the P2 or the synaptosome fractions during standard subcellular fractionation procedure. Although test experiments can be performed to assess the performance of different detergents in terms of preserving the complex integrity [4], it is generally accepted that detergents are likely to disrupt protein binding, especially the weak and transient interactions.

For decades, chemical cross-linking combined with mass spectrometry (XL-MS) has been proposed as a promising alternative to study protein–protein interactions in their physiological context. However, due to the difficulties in detection and identification of cross-linked peptides and the limitations in the depth and sensitivity of such analysis, this approach has been largely restricted in single protein/complex studies. In recent years, with the advancements of highly sensitive and ultra-fast mass spectrometers, in combination with the development of MS-cleavable cross-linkers and data analysis software, proteome-wide XL-MS is emerging as a powerful tool to investigate the structures and interactions of various macromolecular assemblies simultaneously from complex mixtures [6]. In particular, XL-MS workflow is readily applicable to in vivo systems, such as intact organelle and the whole cell [7]. In this aspect, XL-MS allows covalent capturing of the entire interaction network of all membrane proteins in association with their regulatory partners. Thus, it captures a snapshot of functional protein networks in vivo prior to protein extraction under detergent conditions, enabling directly profiling of the native assembly of macromolecular machineries and their cross-talk.

The samples for XL-MS experiments are typically protein mixtures (e.g., single protein/complex or cell lysates), intact organelles (e.g., mitochondria, lysosome), and subcellular fractions (e.g., synaptosome and microsome fractions in our example). During cross-linking reaction, cross-linkers, which are typically a type of small molecules that comprises of two reactive groups connected by a spacer arm, are applied to protein samples to covalently capture the two amino acid residues in close proximity. In the present protocol we use disuccinimidyl sulfoxide(DSSO), a MS-cleavable crosslinker that is reactive towards primary amines (e.g., protein N-termini and lysine side chains). It contains an amine-reactive

N-hydroxy-succinimide ester at each end of a 7-carbon spacer arm that renders a distance constraint of approximately 33.4 Å (23.4 ± 10 Å considering in-solution flexibility) between the two α carbons of the two linked residues. Following cross-linking reaction, cross-linked proteins are subsequently digested into a mixture of peptides using an enzyme of choice, typically trypsin or trypsin in combination with endo-LysC. The resulting peptide mixture contains primarily four major subtypes of peptide products: un-cross-linked peptides, dead-end cross-links (one end of the cross-linker reacts towards proteins and the other end is hydrolyzed in solution), intra-peptide cross-links (the cross-link is within one peptide chain), and inter-peptide cross-links (the cross-link is between two independent peptide chains). Among these four subtypes, inter-peptide cross-links (named as cross-links or cross-linked peptides throughout the chapter) are the most informative ones because they provide distance constraints (i.e. the distance between the α-carbons of two cross-linked residues) in three-dimensional space, which is essential for structural identification and illustration of protein–protein interactions. Cross-links, which are often of lower abundance compared to the large excess of other types of peptides in the mixture, are enriched by either chromatography-based approaches (e.g., strong cation exchange chromatogram and size exclusion chromatogram) or affinity-based approaches (e.g., the use of affinity-tagged cross-linkers).

Lastly, peptide fractions that are enriched for cross-links are subjected to reverse-phase liquid chromatography coupled to mass spectrometry (LC/MS) analysis. Largely benefitted from the unique feature of DSSO cross-linker, which allows a gas-phase fragmentation at the linker spacer arm, the identification of cross-linked peptides has become as simple as a typical linear peptide identification.

The workflow of MS analysis of DSSO cross-linked peptides is detailed in Fig. 1. In essence, the DSSO cross-linker can be prefentially cleaved by collision-induced dissociation (CID) during MS2 fragmentation, generating two pairs of signature peaks with a mass difference of 31.97 Da between the two fragment ions in each pair. If the signature peaks are detected in the CID-MS2 spectrum, an electron transfer dissociation (ETD)-MS2 scan on the same cross-linked precursor and/or targeted MS3 scans on each of the signature ions will be triggered to improve the sequence coverage of the cross-linked peptide. Ultimately, the combined search from data generated from the CID-MS2 scan, the ETD-MS2 scan, and the MS3 scans greatly improve the fidelity of protein identification [8].

While we detailed the protocol to perform the analysis of static protein–protein interaction in synapse, we expect that, in conjunction with TMT tagging (see Chapter 10) XL-MS is applicable to study the dynamic regulation of synaptic protein complexes.

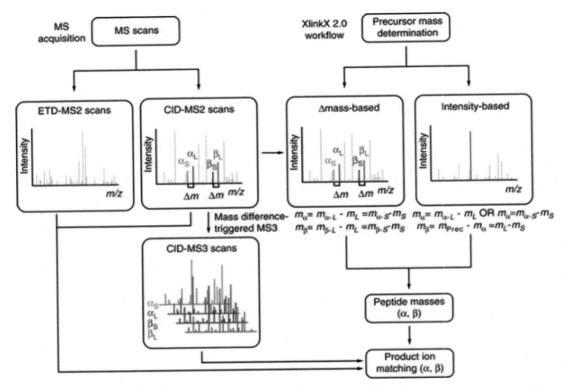

Fig. 1 Each MS1 precursor ion is subjected to sequential CID–MS2 and ETD–MS2 fragmentation. Data-dependent MS3 scans are performed if a unique mass difference (Δm) is found in the CID–MS2 scans. In XlinkX v2.0 data analysis, CID–MS2 scans are used to calculate the potential precursor mass of each linked peptide, using Δm-based, intensity-based, or both strategies. Both CID–MS2 and ETD–MS2 scans, as well as MS3 scans, are subjected to product ion matching to sequence the two peptide constituents of a cross-link. The four signature fragment ions with a unique mass difference (Δm), resulting from CID-induced cross-linker cleavage, are shown in colours

2 Materials

2.1 For Cross-Linking

2.1.1 Stock Solution

500 mM Hepes buffer pH 7.8—Dissolve 11.9 g Hepes in 80 ml distilled water; adjust the pH to 7.8 with NaOH. Fill up to 100 ml with distilled water. Store at 4 °C.

5 M NaCl—Dissolve 29.225 g NaCl in 100 ml distilled water. Store at 4 °C.

1 M Tris–HCl—Dissolve 121 g Tris in 80 ml distilled water, adjusted the pH to 7.5 with HCl. Fill up to 100 ml with distilled water. Store at 4 °C.

2.1.2 Working Buffers

DSSO solution–Dissolve 2 mg DSSO in 50 µl DMSO to make a 100 mM solution. Prepare fresh solution for the cross-link.

Dithiothreitol (DTT).

Chloroacetamide.

8 M urea: dissolve 38.4 g Urea in 3.2 ml 500 mM Hepes, pH 7.4, and fill up to 80 ml with distilled water.

50 mM NH_3HCO_3: dissolve 0.195 g ammonium bicarbonate in 50 ml deionized water.

2.2 Columns and Solvents for Liquid Chromatography

1. Strong cation exchange chromatography:

Column, Zorbax BioSCX–Series II column (0.8 mm inner diameter, 50 mm length, 3.5 μm).

Solvents: solvent A consisted of 0.05% formic acid in 20% acetonitrile, solvent B consisted of 0.05% formic acid, 0.5 M NaCl in 20% acetonitrile.

2. Reversed-phase C18 chromatography:

Column, in-house packed C18 column containing Poroshell 120 EC-C18, 2.7 μm beads from Agilent Technologies.

Solvent A consisted of 0.1% formic acid, solvent B consisted of 0.1% formic acid in 80% acetonitrile.

2.3 Liquid Chromatography Systems and Mass Spectrometer

1. SCX HPLC system:

A HPCL system that can deliver flow rates in the range of 100 μl/min–1 ml/min. We use an ultra HPLC Agilent 1200 system (Agilent Technologies).

2. LC-MS/MS system:

The preferred LC/MS system is nano-LC sytems coupled to Orbitrap Fusion/Lumos mass spectrometers (Thermo Fisher Scientific). We use an ultra-HPLC Proxeon EASY-nLC 1000 system (Thermo Fisher Scientific) coupled to an Orbitrap Fusion mass spectrometer.

3 Methods

3.1 Cross-Linking of Proteins in Synaptosome Fraction

1. Prepare synaptosomes as described in detail in Chapter 3.

2. Resuspend the synaptosomes in resuspension buffer (0.32 M sucrose, 5 mM Hepes, pH 7.8,) to a final protein concentration of 1 mg/ml, and transfer to a 1.5 ml Eppendrof tube (*see* **Note 1**).

3. Add 100 mM DSSO stock solution to the synaptosome sample to reach a final concentration of 1 mM DSSO. For example, 10 μl 100 mM DSSO is added to 1 ml synaptosome sample.

4. Incubate the reaction mixture for 45 min at room temperature with gentle shaking.

5. Add Tris–HCl 1 M buffer to the cross-linked synaptosome to reach a final concentration of 10 mM (100 times dilution) and incubate for 30 min to quench the reaction.

6. Centrifuge the sample in a table centrifuge at maximum speed for 15 min to pellet the synaptosome.

3.2 Digestion of Proteins

1. Resuspend cross-linked synaptosome pellet in 250 μl freshly prepared lysis buffer (8 M urea in 20 mM Hepes, pH 7.8).

2. Add DTT to a final concentration of 5 mM. Incubate at 56 °C for 30 min.

3. Add chloroacetamide to a final concentration of 40 mM (see Note 2). Incubate for 30 min in the dark.

4. Digest the sample with Lys-C at an enzyme-to-protein ratio of 1:75 (wt/wt) at 37 °C for 4 h.

5. Dilute the sample fourfold in 50 mM ammonium bicarbonate. Mix well.

6. Add trypsin at an enzyme-to-protein ratio of 1:100 (wt/wt). Incubate at 37 °C overnight.

7. Desalt the peptide mixture using Sep-Pak C18 cartridges (Waters Corporation).

8. Dry the eluted peptides in speedvac. Store at −20 °C for further use.

3.3 LC and LC-MS/MS Analysis

1. Re-dissolve the sample in 20 μl 10% formic acid, vortex for 10 s to fully suspend the peptide.

2. Load the sample onto a strong cation exchange column. Elute the peptides from the column with the following gradient; 0–0.01 min (0–2% B); 0.01–8.01 min (2–3% B); 8.01–14.01 min (3–8% B); 14.01–28 min (8–20% B); 28–38 min (20–40% B); 38–48 min (40–90% B); 48–54 min (90% B); and 54–60 min (0% B). Collect 50 fractions for each sample, dry, and store at −20 °C. We use fractions 20–40 for subsequent LC-MS analysis. Peptides contained in these factions are predominately longer and higher charged with $MH^+ > 3$, which are characteristic of cross-linked peptides.

3. Dissolve each fraction in 10 μl 5% actronitrile and 0.1% formic acid and inject 2 μl onto the nano-LC system. The flow rate is set to 100 nl/min. We use a 180-min gradient (7–30% solvent B within 151 min, 30–100% solvent B within 3 min, 100% solvent B for 5 min, 100–7% solvent B within 1 min and 7% solvent B for 20 min).

4. For the MS scans, set the scan range to 350–1500 m/z at a resolution of 120,000, and the automatic gain control (AGC) target at 1×10^6. For the MS/MS scans, set the resolution to 60,000, the AGC target to 1×10^5, the precursor isolation width 1.6 Da, the maximum injection time 120 ms, and the

CID normalized collision energy at 25%. MS3 spectra are acquired in the ion trap mass analyser when peak doublets with a specific mass difference (Δ = 31.9721 Da) is detected in the CID-MS2 spectrum. This mass difference is indicative for the presence of DSSO cross-linked peptides.

3.4 Analysis of Cross-Link Data

1. Use Protein Discoverer to convert the peak list from each RAW file into three MGF files containing the CID-MS2, ETD-MS2 and CID-MS3 data, respectively.

2. Deconvolute the CID- and ETD-MS2 spectra to charge state 1 using the MS2 Spectrum Processor add-on node in Proteome discoverer v1.4.

3. The MGF files were used as input to identify cross-linked peptides with standalone XlinkX v2.0.

4. Operate XlinkX with these settings: MS ion mass tolerance: 10 ppm; MS2 ion mass tolerance: 20 ppm; MS3 ion mass tolerance, 0.6 Da; fixed modification: Cys carbamidomethylation; variable modification: Met oxidation; enzymatic digestion: trypsin; allowed number of missed cleavages: 3.

5. Search the MS2 and MS3 spectra against the Swiss-Prot database of *Mus musculus* proteins.

4 Notes

1. Cross-linking buffer should not contain amine groups in order to avoid quenching of the cross-linker (e.g. do not use Tris containing buffers).

2. DTT and chloroacetamide are the reagents for reduction of disulfide bridges and alkylation of the free SH-groups, respectively. The other commonly used reagents can also be used [9]. For example the reducing tris-2(-carboxyethyl)-phosphine (TCEP) and β-mercaptoethanol; and the alkylation reagents including iodoacetic acid, iodoacetamide, acrylamide (*see* Chapter 12 from first edition of Neuroproteomics [10]), *N*-ethylmaleimide, methyl methanethiosulfonate (MMTS), and 4-vinylpyridine.

References

1. Chua JJ, Kindler S, Boyken J, Jahn R (2010) The architecture of an excitatory synapse. J Cell Sci 123:819–823

2. Pandya NJ, Koopmans F, Slotman JA, Paliukhovich I, Houtsmuller AB, Smit AB, Li KW (2017) Correlation profiling of brain subcellular proteomes reveals co-assembly of synaptic proteins and subcellular distribution. Sci Rep 7:12107

3. Koopmans F, Ho JTC, Smit AB, Li KW (2018) Comparative analyses of data independent acquisition mass spectrometric approaches: DIA, WiSIM-DIA, and untargeted DIA. Proteomics 18:1700304

4. Chen N, Pandya NJ, Koopmans F, Castelo-Szekelv V, van der Schors RC, Smit AB, Li KW (2014) Interaction proteomics reveals brain region-specific AMPA receptor complexes. J Proteome Res 13:5695–5706

5. Pandya NJ, Klaassen RV, van der Schors RC, Slotman JA, Houtsmuller A, Smit AB, Li KW (2016) Group 1 metabotropic glutamate receptors 1 and 5 form a protein complex in mouse hippocampus and cortex. Proteomics 16:2698–2705

6. Liu F, Heck AJ (2015) Interrogating the architecture of protein assemblies and protein interaction networks by cross-linking mass spectrometry. Curr Opin Struct Biol 35:100–108

7. Liu F, Lossl P, Rabbitts BM, Balaban RS, Heck AJR (2018) The interactome of intact mitochondria by cross-linking mass spectrometry provides evidence for coexisting respiratory supercomplexes. Mol Cell Proteomics 17:216–232

8. Liu F, Lossl P, Scheltema R, Viner R, Heck AJR (2017) Optimized fragmentation schemes and data analysis strategies for proteome-wide cross-link identification. Nat Commun 8:15473

9. Muller T, Winter D (2017) Systematic evaluation of protein reduction and alkylation reveals massive unspecific side effects by iodine-containing reagents. Mol Cell Proteomics 16:1173–1187

10. Chen N, Schors RC v, Smit AB (2011) A 1D-PAGE/LC-ESI linear ion trap orbitrap MS approach for the analysis of synapse proteomes and synaptic protein complexes. In: Li KW (ed) Neuroproteomics. Humana Press, Totowa, NJ, pp 159–167

Chapter 10

TMT-MS³-Enabled Proteomic Quantification of Human IPSC-Derived Neurons

Nikhil J. Pandya, David Avila, Tom Dunkley, Ravi Jagasia, and Manuel Tzouros

Abstract

The induced pluripotent stem cell (IPSC)-derived neurons technology has matured to a point where neurological disorders such as autism, schizophrenia, Alzheimer's, and Parkinson's disease can be reproducibly modeled. Proteomic analysis of these model systems has the potential to identify disease signatures, characterize treatment effects, and drive biomarker identification. Here we describe the implementation of a TMT-MS³-based proteomic workflow for the characterization of over 7000 proteins in whole cell lysates obtained from human IPSC-derived neurons. This multiplexing protocol should have a better reproducibility and higher number of assessed conditions than the label-free or SILAC-based proteomics approaches.

Key words IPSC-derived neurons, TMT, MS³, Proteomics, Orbitrap Lumos, Synchronous Precursor Selection

1 Introduction

1.1 Modeling Neuropsychiatric Disorders Using Human-Induced Pluripotent Stem Cells Technology

The advent of the stem cell technology has allowed the establishment of efficient reprograming strategies to convert somatic cells, such as adult fibroblasts, into induced pluripotent stem cells, which can then be effectively differentiated into any other cell type [1]. Following this, there has been a surge in studies which has allowed the application to samples obtained directly from humans. In particular, protocols for the reprograming of previously inaccessible tissues, such as neurons, have tremendous potential in disease modeling of neuropsychiatric disorders.

The conversion of somatic cells into induced pluripotent stem cells (IPSCs) involves using reprograming factors [1] to obtain a population of self-renewing expandable population of neural progenitor cells (NPCs), which can be further differentiated into neurons of various subtypes [2–4]. This allows not only to capture the

Ka Wan Li (ed.), *Neuroproteomics*, Neuromethods, vol. 146,
https://doi.org/10.1007/978-1-4939-9662-9_10, © Springer Science+Business Media, LLC, part of Springer Nature 2019

effects of highly penetrant genetic aberrations for monogenic disorders in a heterogeneous background but also to model a large set of polygenic disorders, involving multiple alleles with relatively small effect sizes. To this end, a large variety of studies have now focused on modeling neuropsychiatric disorders using stem cell technologies (reviewed in [5–7]). Since several of these disorders are caused by aberrant functioning of proteins, proteomics has enormous potential to capture the underlying molecular pathophysiology of neurons obtained from patients, test the effects of highly penetrant mutations in an isogenic control line, and even test reversibility of pathophysiology for drug screening (Fig. 1) at various time points during differentiation. With recent advances in sample preparation, mass spectrometry technology and the development of labeling strategies for quantification, rapid, sensitive, and comprehensive interrogation of the proteomic changes in IPSC-derived neurons is now a reality.

1.2 Labeling Strategies for Quantitative Proteomics

MS-based proteomics has the potential for identifying over 10,000 proteins from complex mixtures, such as whole cell lysates, in just a few hours of liquid chromatography-mass spectrometry (LC-MS) time [8]. Several proteomic strategies are available for protein quantification [9]. Among these, "label-free" proteomics approaches are known to be less accurate than ones relying on met-

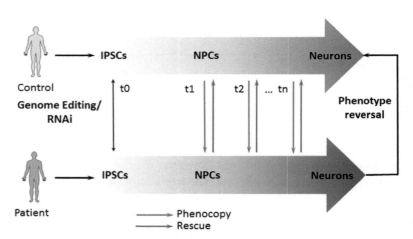

Fig. 1 Modeling neuropsychiatric disorders using IPSC-derived neurons. IPSCs from healthy control samples can be obtained and compared using proteomics to patient derived cells at all stages of differentiation (IPSCs, NPCs, neurons) to identify key elements driving the molecular pathophysiology in disease. Reversibility of the phenotypes can be tested at various time points by either generating isogenic lines, enabling testing of rescue strategies in a time-dependent manner (t_1, t_2, ..., t_n)

abolic or chemical labeling. Importantly, as the number of samples that need to be measured expands, "label-free" methods suffer from long measurement times, leading to high dependence on stable chromatography and measurements. In order to circumvent some of these issues, quantitative approaches based on metabolic labeling, such as stable isotope labeling with amino acids in cell culture (SILAC), have been developed [10]. Besides its undisputed utility for quantitation, one limitation of the SILAC technology is that it allows comparison of at most three conditions. The cost of pooling conditions in the SILAC approach is increased sample complexity and consequently a reduction in protein identification rates. An alternative approach is chemical labeling using isobaric tags such as tandem mass tags (TMT) or isobaric tags for relative and absolute quantification (iTRAQ). These methods promise efficient multiplexing and deep proteome coverage for up to 11 and 8 samples, for TMT and iTRAQ, respectively. Peptides labeled with these reagents have indistinguishable precursor masses at the MS^1 level and consequently multiplexing does not impact the complexity of the mixture. After fragmentation, peptides from each sample give rise to a reporter ion, enabling relative quantification [11]. Deep coverage can be obtained using fractionation and since the samples are labeled, pooled and measured together, these methods are less sensitive to changes in LC-MS conditions, compared to "label-free" approaches.

One inherent problem of TMT and iTRAQ labeling is ratio-compression due to co-isolation and co-fragmentation of multiple peptide ions (*see* Fig. 2). Since the isobaric tags originating from peptides of interest are indistinguishable from interfering peptides, the final reporter ion intensities can be unpredictably contaminated from co-isolated peptides [12, 13]. In order to prevent this effect, an MS^3-based method was developed. However this approach led to loss of sensitivity and reduced quantification depth [14]. To circumvent these issues, a method, initially termed multi-notch MS^3, was established, which involved selection and co-fragmentation of multiple MS^2 ions [15]. This dramatically reduces the impact of interference on the reporter ion signature without significantly affecting sensitivity. The latter method is now commercially implemented on Orbitrap tribrid systems and called synchronous precursor selection (SPS) (Fig. 2).

In this chapter, we demonstrate an optimized sample preparation and analysis strategy (Fig. 3) that can be applied to cell pellets obtained from a single well of a 6 well plate to robustly quantify over 7000 proteins using a TMT-MS³ strategy.

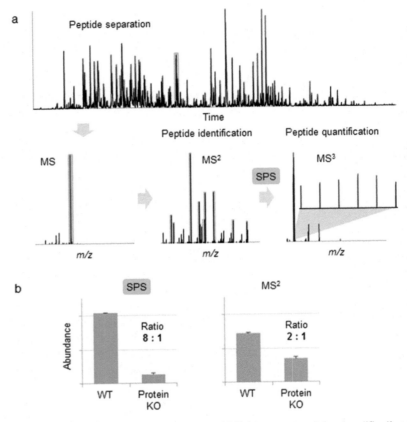

Fig. 2 Synchronous precursor selection (SPS) improves protein quantification accuracy. (**a**) Peptides from a mixture are first separated by reversed phase LC and their mass-to-charge (*m/z*) measured by MS. A precursor is then selected for fragmentation by collision-induced dissociation (CID) to generate an MS^2 used for identification. In a next step, several peptide fragment ions (10 in the example, marked in blue) are co-isolated and fragmented by SPS using high-energy collision dissociation (HCD) MS^3 to liberate the TMT reporter ions used for quantification. (**b**) Effect of SPS vs. MS^2 on the quantification accuracy of a protein knockout (KO) experiment. Using MS^2, the measured ratio is only 2:1 (compared to 8:1 using SPS) due to "ratio-compression"

2 Materials

2.1 Lysis and Sample Preparation

1. Tip Sonifier: QSonica, Newtown, CT.
2. Solid-phase extraction columns (50 mg C18 Sep-Pak, Waters, Milford, MA).
3. Lyophilizer: FreeZone 2.5 Plus, Labconco, Kansas City, MO.

2.2 Offline High-pH Reversed-Phase Fractionation

1. Agilent 1260 infinity series HPLC (Agilent Technologies, Waldbronn, Germany).

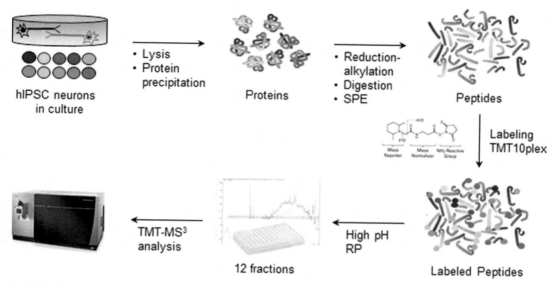

Fig. 3 Workflow for TMT labeling and subsequent SPS analysis. Cells in culture are lysed to obtain proteins, which are subsequently precipitated, reduced, alkylated and digested followed by solid phase extraction (SPE). The resulting peptides from each sample are labeled with TMT10 plex reagents, pooled and separated by high-pH reversed-phase (RP) fractionation followed by TMT-MS³ analysis

2. YMC-Triart C18 Column (0.5 mm × 250 mm, S-3 μm particle size, 12 nm pore size).

3. 96 well Sample plate (Thermo Fisher Scientific, Rockford, IL, USA).

2.3 Synchronous Precursor Selection (SPS) Acquisition

1. Orbitrap Fusion Lumos Tribrid (Thermo Fisher Scientific) mass spectrometer.

2.4 Software

1. Proteome Discoverer 2.1 (Thermo Fisher Scientific) or later versions.

2.5 Reagents

Note: water for all buffer preparation is of Milli-Q (Millipore) quality.

1. Lysis buffer: 1% (w/v) SDS, 8 M urea, 50 mM ammonium bicarbonate pH 8.8, protease inhibitor cocktail EDTA-free (Roche), phosphatase inhibitor cocktail (Roche).

2. Chloroform-methanol precipitation: chloroform HPLC grade (Sigma), methanol HPLC grade (Sigma).

3. Resuspension buffer: 8 M urea, 50 mM ammonium bicarbonate pH 8.8.

4. TMT10plex isobaric labeling reagent (Thermo Fisher Scientific, Rockford, IL, USA).

5. Protein concentration estimation: bicinchoninic assay (BCA assay kit, Pierce).

6. Digestion reagents: lysyl endopeptidase (Lys-C, Wako Pure Chemical Industries Ltd., Japan), trypsin (Promega, Madison, WI, USA).

7. Reduction reagent: 100 mM dithiothreitol (DTT, Sigma).

8. Alkylating reagent: 200 mM iodoacetamide (IAA, Sigma).

9. Solvents for SPE: solvent A: acetonitrile (ACN) (HPLC grade, Sigma); solvent B: 50% (v/v) ACN; wash solvent: 5% (v/v) formic acid (FA) (HPLC grade, Sigma).

10. Solvents for high-pH reversed-phase fractionation: solvent A: 10 mM ammonium formate pH 10.0; solvent B: 10% (v/v) solvent A in ACN.

11. Solvents for LC-MS acquisition: solvent A: 0.1% (v/v) FA; solvent B: 0.1% (v/v) FA in ACN.

3 Methods

3.1 Lysis

Note: All steps are done at room temperature unless stated otherwise.

1. Cell pellets from 1 well of a six well plate (~1 million cells) stored at −80 °C are transferred to room temperature immediately after the addition of 150 μl of lysis buffer and mixed by pipetting up and down ten times until the cell pellet is dissolved.

2. The cell lysates are sonicated to remove undissolved particles and shear DNA by setting power at 30%, for 3 × 10 cycles (1 s ON, 1 s OFF) while keeping the samples on ice.

3. Lysates are then subjected to methanol-chloroform protein precipitation [16]. The volumes can be scaled linearly based on the volume of the sample.

   (a) Mix the cell lysate (150 μl) with 600 μl of methanol and vortex for 10 s.

   (b) Add 150 μl of chloroform and vortex again until the solution is homogenous.

   (c) Add 450 μl of water and vortex again for 30 s. The solution should turn white.

   (d) Centrifuge at 20,000 × g for 15 min.

   (e) Remove the upper organic phase with a pipette without disturbing the white protein pellet at the interface.

   (f) Add 450 μl of cold methanol and vortex for 30 s.

   (g) Centrifuge at 20,000 × g for 5 min.

(h) Discard the supernatant without disturbing the protein pellet and repeat the methanol wash twice.

(i) Air-dry the protein pellet for approximately 15 min.

(j) Dissolve the pellet in 50 μl of resuspension buffer by shaking for 15 min at 1000 rpm (agitation frequency). Eventually use a bath sonicator to aid pellet dissolution.

4. Perform protein quantification using the BCA assay according to manufacturer's instructions (Pierce).

3.2 Reduction, Alkylation and Protein Digestion

1. Proteins are reduced using 5 μl of reducing agent for 30 min at 56 °C with shaking at 1000 rpm.

2. The samples are cooled down to room temperature and alkylated using 5 μl freshly prepared alkylating reagent for 30 min with shaking at 1000 rpm in dark.

3. Samples are predigested by adding lysyl endopeptidase (Lys-C) dissolved in water at a ratio of 1:100 (w/w protein) for 4 h at 37 °C with shaking at 1000 rpm.

4. Samples are diluted fourfold with 50 mM ammonium bicarbonate and digested with trypsin solution with final ratio of 1:100 (w/w protein) (100 μg trypsin in 500 μl of 50 mM acetic acid; 0.2 μg/μl) overnight at 37 °C with shaking at 1000 rpm.

3.3 Solid Phase Extraction

1. The samples are acidified using neat FA to a final concentration of 5% (v/v), vortexed and kept on ice for 10 min.

2. Centrifuge the samples at 10,000 × g for 5 min. (Note: The white insoluble pellet formed at the bottom consists mainly of lipids). Collect the supernatant and transfer to a new tube.

3. Wet the C18 column using 500 μl solvent A.

4. Wash the column with 1 ml solvent B.

5. Equilibrate the column with 2 ml wash solvent.

6. Load the sample on the C18 column.

7. Wash the column with 2 ml wash solvent.

8. Wash column with 1 ml water.

9. Elute the sample in a new collection tube with 250 μl solvent B.

10. Transfer the volume corresponding to 30 μg protein/peptide to a new tube, lyophilize overnight, and store the lyophilized peptides at −20 °C.

3.4 Labeling Using TMT Reagent

1. 30 μg lyophilized peptides are reconstituted in 70 μl of 100 mM HEPES pH 8.5 buffer.

(*Caution*: check pH of the samples to ensure that the samples are ~pH 8–9).

2. Dissolve TMT10plex (0.8 mg) reagent in 130 μl of anhydrous ACN.

3. Add 30 μl of dissolved TMT10plex reagent to the samples (1:6 w/w peptide to reagent) and incubate with shaking at 1000 rpm.

4. Labeling reaction is quenched using a 5% (v/v) hydroxylamine solution (0.5% v/v final concentration) for 15 min with shaking at 1000 rpm.

5. The samples for a single TMT10plex experiment are combined into aliquots of 150 μg total protein/peptide and acidified using FA (5% v/v final concentration). An aliquot corresponding to 5 μg total protein/peptide is set aside for labeling efficiency test.

6. The pooled TMT10plex samples are cleaned up using the SPE procedure and lyophilized overnight.

3.5 Labeling Efficiency Test

1. 5 μg of saved sample is dissolved in dissolution solvent (5% v/v FA, 2% v/v ACN).

2. Samples are run in data-dependent acquisition mode (*see* parameters for DDA).

3. Raw data are processed using Proteome Discoverer 2.1 using parameters mentioned in the labeling efficiency test PD parameters and TMT10 as a dynamic modification.

4. Ensure that the labeling efficiency is >95% (TMT modified versus unmodified PSMs).

3.6 Offline High-pH Reversed-Phase Fractionation

1. Dissolve 150 μg pooled sample in 98% (v/v) solvent A 2% (v/v) solvent B by shaking at 1000 rpm.

2. Centrifuge at $10,000 \times g$ for 10 min and transfer supernatant into an autosampler vial.

3. Fractionate the samples using the following gradient at 12 μl/min (Fig. 4).

   (a) 2–23% solvent B for 5 min, 23–33% solvent B for 25 min, 33–53% solvent B for 30 min, 53–100% solvent B for 5 min and 100% solvent B for 5 min. The column is equilibrated by changing from 100% solvent B to 2% solvent B in 1 min followed by 2% solvent B for 14 min.

   (b) A total of 36 fractions are collected in a 96 well sample plate from 4 to 84 min consisting of ~26 μl volume each.

   (c) Each 3 fractions from a row are combined together (fractions 1, 13, and 25; fractions 2, 14, and 26; ...; fractions 12, 24, and 36) to obtain 12 pooled fractions.

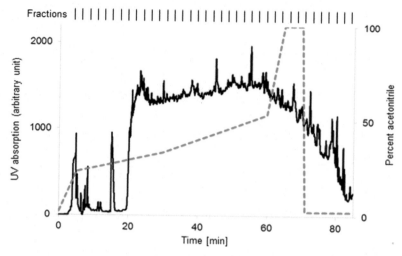

Fig. 4 Example UV trace for 150 μg IPSC-derived neurons digest sample using offline high-pH reversed-phase fractionation. Fraction collection tick marks are indicated

4. Acidify samples by adding formic acid (5% v/v final concentration).

5. Samples (~12 μg per pooled fraction) are then split into aliquots of ~4 μg each, dried in a vacuum concentrator and stored at −20 °C until measurement.

3.7 TMT-MS³ Analysis Using an Orbitrap Fusion Lumos

1. Dissolve fractionated samples with 40 μl 5% formic acid, 2% acetonitrile and inject 1.5 μg peptide equivalent for LC-MS by loading on an Acclaim PepMap C18 trapping column (100 μm × 20 mm, 5 μm particle size) at a controlled maximum backpressure of 500 bar.

2. Peptides are then separated on Acclaim PepMap C18 EASY-spray column (75 μm × 750 mm, 2 μm particle size) heated at 50 °C with the following gradient at 270 nl/min.

(a) 5% solvent B for 5 min, 5–20% solvent B for 120 min, 20–45% solvent B for 90 min, 45–100% solvent B for 5 min, 100% solvent B for 20 min 5–15% solvent B.

3. Acquisition parameters:

(a) The instrument is operated in data-dependent acquisition mode to collect Orbitrap MS¹ scans over a mass range of 350–1400 *m/z* at a resolution of 120,000 (at *m/z* 200) with an automatic gain control (AGC) target value of 2E5 with maximum injection time (IT) of 50 ms.

(b) Data is calibrated on the fly using ambient air hexacyclodimethylsiloxane at *m/z* 445.12002.

(c) Between each MS^1 scan, for a period of 3 s, the N most intense precursor ions with charge states between 2 and 6, with a minimum intensity of 5E3, are mono-isotopically selected for collision-induced dissociation (CID), using a quadrupole isolation of m/z 0.7, AGC target 1E4, maximum IT 50 ms, collision energy of 35%, and ion trap readout with turbo scan rate.

(d) Precursor ions are excluded after 1 appearance for 75 s using 10 ppm as low and high mass tolerances. The dependent scan is performed on a single charge state per precursor.

(e) TMT reporter ions are generated using synchronous precursor selection (SPS), an MS quadrupole isolation window of m/z 2, high-energy collision dissociation (HCD) at a normalized collision energy of 65%, and readout in the Orbitrap with a resolution of 50k (at m/z 200), scan range of m/z 100–500, an AGC target of 5E4, and a maximum IT of 105 ms. The mass range for selecting the SPS precursors is from m/z 400 to 2000, excluding the MS^2 precursor with a tolerance of m/z 40 (low) and 5 (high), and any TMT neutral loss from it. The number of SPS precursors is set to 10.

3.8 Data Processing in Proteome Discoverer

1. Post-acquisition, the raw data is processed using Proteome Discoverer 2.1 connected to Mascot Server 2.6.1 (Matrix Science, London, UK).

2. Processing workflow Parameters: See PD workflow parameters for detailed workflow.

   (a) The processing workflow searches the MS^2 data against the UniProt human protein database using trypsin/P as an enzyme, allowing for a maximum of 2 missed cleavages and 10 ppm and 0.5 Da precursor and fragment ion tolerances, respectively.

   (b) Carbamidomethylated cysteine (+57.02146 Da), TMT10 labeled lysine and peptide N-terminus (+229.16293 Da) are set as static modifications.

   (c) Oxidized methionine (+15.99492 Da) and acetylated protein N-terminus (+42.01057 Da) are set as dynamic modifications.

   (d) A decoy database search is performed using Percolator with the Target FDR set to 0.01 based on q-value threshold.

(e) Reporter ion quantification is performed on the HCD-MS³ data, with 3 mmu peak integration and using the most confident centroid tolerances.

(f) Reporter ion intensities are adjusted so as to correct for the isotope impurities of different TMT reagents using the manufacturer specifications.

3. Consensus workflow Parameters: See PD workflow parameters for detailed workflow.

(a) A consensus workflow is defined to group PSMs into peptide and proteins.

(b) Peptide FDRs are controlled by setting a q-value threshold of 0.01 and allowing the software to automatically select PSM q-value for the grouping.

(c) High-confidence unique peptides with a minimal length of 6 amino acids are grouped into proteins and protein FDR is also set to 0.01.

(d) Peptide and protein quantification is done by summing the S/N for each channel and normalizing each value with the highest TMT channel total intensity. Individual peptide and protein S/N are scaled to an average of 100 and only high-FDR-confidence protein quantification intensities are kept for statistical analysis.

3.8.1 Proteome Discoverer 2.1 Processing Workflow

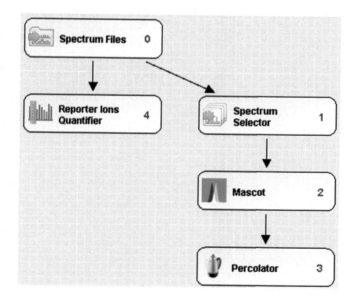

Node name (WF node #)	Parameter cat.	Parameter	Setting or value
Spectrum selector (**1**)	3. Scan event filters	MS order	Is MS^2
Mascot (**2**)	1. Input data	Protein database	SwissProt
		Enzyme name	Trypsin/P
		Max. miss. cleav. sites	2
		Instrument	ESI-Trap
		Taxonomy	*Homo sapiens* (human)
	2. Tolerances	Prec. mass tolerance	10 ppm
		Frag. mass tolerance	0.5 Da
	4. Dynamic modifications	1. Dynamic modification	Oxidation (M)
		2. Dynamic modification	Acetyl (protein N-term)
	5. Static modifications	1. Static modification	Carbamidomethyl (C)
		2. Static modification	TMT10plex (N-term)
		3. Static modification	TMT10plex (K)
Percolator (**3**)	2. Decoy database search	Target FDR (strict)	0.01
		Validation based on	*q*-Value
Reporter ions quant. (**4**)	1. Peak integration	Integration tolerance	3 mmu
		Integration method	Most confident centroid
	2. Scan Event filters	Mass analyzer	FTMS
		MS order	MS^3
		Activation type	HCD

3.8.2 Proteome Discoverer 2.1 Processing Workflow

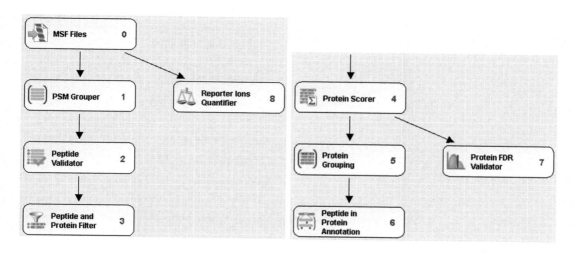

Node name (WF node #)	Parameter Cat.	Parameter	Setting or value
MSF files (**0**)	1. Spectra storage set.	Spectra to store	Identified or quant.
	4. PSM filters	Max. rank	0
PSM grouper (**1**)	1. Peptide group mod.	Site prob. threshold	25
Peptide validator (**2**)	1. General validation settings	Validation mode	Automatic (control peptide level error rate if possible)
		Target FDR (strict) for PSMs	0.01
		Target FDR (strict) for peptides	0.01
Peptide and protein filter (**3**)	1. Peptide filters	Peptide conf. at least	High
		Min. peptide length	6
Protein grouping (**5**)	1. Protein grouping	Apply strict parsimony principle	True
Peptide in protein annotation (**6**)	2. Mod. in peptide	Protein mod. reported	For all proteins
	3. Mod. in protein	Mod. sites reported	All combined
	4. Positions in protein	Protein pos. for peptides	For all proteins
Protein FDR validator (**7**)	1. Confidence thresholds	Target FDR (strict)	0.01 (protein)

(continued)

Node name (WF node #)	Parameter Cat.	Parameter	Setting or value
Peptide and protein quantifier (**8**)	1. Quantification—general	Peptides to use	Unique
		Consider protein groups for peptide uniqueness	True
		Reject quan results with missing channels	False
	2. Reporter quantification	Reporter abundance based on	S/N
		Apply quan. value corr.	True
		Co-isolation threshold	50
		Average reporter S/N threshold	10
	4. Normalization and scaling	Normalization mode	Tot. pep. amount
		Scaling mode	On all average

References

1. Takahashi K, Tanabe K, Ohnuki M, Narita M, Ichisaka T, Tomoda K, Yamanaka S (2007) Induction of pluripotent stem cells from adult human fibroblasts by defined factors. Cell 107(5):861–872

2. Shi Y, Kirwan P, Smith J, Robinson HPC, Livesey FJ (2012) Human cerebral cortex development from pluripotent stem cells to functional excitatory synapses. Nat Neurosci 15(3):477–486

3. Kriks S, Shim J-W, Piao J, Ganat YM, Wakeman DR, Xie Z, Carrillo-Reid L, Auyeung G, Antonacci C, Buch A, Yang L, Beal MF, Surmeier DJ, Kordower JH, Tabar V, Studer L (2011) Dopamine neurons derived from human ES cells efficiently engraft in animal models of Parkinson's disease. Nature 480(7378):547–551

4. Sances S, Bruijn LI, Chandran S, Eggan K, Ho R, Klim JR, Livesey MR, Lowry E, Macklis JD, Rushton D, Sadegh C, Sareen D, Wichterle H, Zhang S-C, Svendsen CN (2016) Modeling ALS with motor neurons derived from human induced pluripotent stem cells. Nat Neurosci 16(4):542–553

5. Adegbola A, Bury LA, Fu C, Zhang M, Wynshaw-Boris A (2017) Concise review: induced pluripotent stem cell models for neuropsychiatric diseases. Stem Cells Transl Med 6(12):2062–2070

6. Liu Y, Deng W (2016) Reverse engineering human neurodegenerative disease using pluripotent stem cell technology. Brain Res 1638:30–41

7. Haggarty SJ, Silva MC, Cross A, Brandon NJ, Perlis RH (2016) Advancing drug discovery for neuropsychiatric disorders using patient-specific stem cell models. Mol Cell Neurosci 73:104–115

8. Aebersold R, Mann M (2016) Mass-spectrometric exploration of proteome structure and function. Nature 537(7620):347–355

9. Bantscheff M, Lemeer S, Savitski MM, Kuster B (2012) Quantitative mass spectrometry in proteomics: Critical review update from 2007 to the present. Anal Bioanal Chem 404(4):939–965

10. Ong SE (2012) The expanding field of SILAC. Anal Bioanal Chem 404(4):967–976

11. Rauniyar N, Yates JR (2014) Isobaric labeling-based relative quantification in shotgun proteomics. J Proteome Res 13(12):5293–5309

12. Karp NA, Huber W, Sadowski PG, Charles PD, Hester SV, Lilley KS (2010) Addressing accuracy and precision issues in iTRAQ quantitation. Mol Cell Proteomics 9(9):1885–1897

13. Ow SY, Salim M, Noirel J, Evans C, Rehman I, Wright PC (2009) iTRAQ underestimation in simple and complex mixtures: the good, the

bad and the ugly. J Proteome Res 8(11):5347–5355

14. Wühr M, Haas W, McAlister GC, Peshkin L, Rad R, Kirschner MW, Gygi SP (2012) Accurate multiplexed proteomics at the MS² level using the complement reporter ion cluster. Anal Chem 84(21):9214–9221

15. McAlister GC, Nusinow DP, Jedrychowski MP, Wühr M, Huttlin EL, Erickson BK, Rad R, Haas W, Gygi SP (2014) MultiNotch MS³ enables accurate, sensitive, and multiplexed detection of differential expression across cancer cell line proteomes. Anal Chem 86(14):7150–7158

16. Wessel D, Flügge UI (1984) A method for the quantitative recovery of protein in dilute solution in the presence of detergents and lipids. Anal Biochem 138(1):141–143

Chapter 11

Data-Independent Acquisition (SWATH) Mass Spectrometry Analysis of Protein Content in Primary Neuronal Cultures

Miguel A. Gonzalez-Lozano and Frank Koopmans

Abstract

Sequential Window Acquisition of all THeoretical fragment-ion (SWATH) is a recently developed discovery proteomics technique based on Data-Independent Acquisition (DIA) mass spectrometry. In this approach, MS/MS is performed simultaneously on all peptides contained in a predefined wide-open mass window of up to 25 Da. The mass window is sequentially stepped through over the entire mass range, usually between 400 and 1200 Da that covers most peptides. As quantitative MS/MS information is generated for all observable peptides in the sample, the missing data and variability between replicates are substantially reduced when compared to a Data-Dependent Acquisition approach. To identify each peptide from the high complexity of the MS/MS spectra generated from multiple peptides, a comprehensive reference spectral library derived prior from a similar sample by Data-Dependent Acquisition, rather than a conventional genome-wide database, should be used.

In this chapter, we describe a general protocol that benefits from the advances of SWATH-MS for the quantification of the primary neuronal culture proteome.

Key words SWATH-MS, DIA, Primary culture, Neuron, Quantitative proteomics

1 Introduction

Bottom-up quantitative proteomics based on LC-MS/MS is the preferential method to measure protein expression levels in biological samples, including the nervous system [1]. Proteins are extracted from the biological source and enzymatically digested into peptides, which are then separated by liquid chromatography and analyzed by mass spectrometry. Traditionally, data-dependent acquisition (DDA) has been the most widely used approach for discovery proteomics [2]. Using this strategy, a full mass MS1 scan is followed by the sequential selection of the top 5 to 40 most intense precursor ions (depending on the type of mass spectrometer) for a subsequent MS/MS scan. The fragment ion spectra from MS/MS scans are matched to their corresponding sequences allowing the identification of tens of thousands of peptides in an

Ka Wan Li (ed.), *Neuroproteomics*, Neuromethods, vol. 146,
https://doi.org/10.1007/978-1-4939-9662-9_11, © Springer Science+Business Media, LLC, part of Springer Nature 2019

experiment. However, the stochastic nature of precursor ion selection can compromise reproducibility, especially for complex samples and low abundant peptides. The missing peptide identifications among replicates (up to 30% [3]) strongly reduce the number of identified proteins.

SWATH mass spectrometry (a type of Data-Independent Acquisition, DIA) has recently emerged as an alternative to overcome the previous limitations [4]. In this mode, all peptides within a predefined mass window ranged from 3 to 25 Da are fragmented and analyzed by MS/MS. The mass windows are sequentially stepped through in ranges usually between 400 and 1200 Da, which should cover the majority of peptides [5]. However, the MS/MS spectrum of each window is highly complex because it includes fragment ions from multiple peptides. Data analysis with a traditional DDA database search strategy, originally designed for the identification of a single peptide per MS/MS scan, is not optimal. A way to bypass this obstacle is the generation of a reference spectral library beforehand by DDA [6, 8]. The information of the elution time and fragment ions of the peptides can be used to link back each fragmentation pattern in the DIA set to the corresponding precursor. Recently, new strategies have been developed to enable the analysis without the need of extra information from a reference library [7, 9], at an expense of a lower number of identified peptides [5].

SWATH has been shown to be superior to DDA for label-free quantitative analysis [10]; a higher number of proteins can be quantified in shorter mass spectrometer analysis time with fewer missing values. The coefficient of variation across replicates is lower, making it especially suitable for capturing subtle protein expression changes between groups or conditions. In addition, raw data can be re-interrogated in silico afterwards to test a new hypothesis (using another dedicated spectral library in accordance with the hypothesis) without repeating the experiments [11].

The advantages granted by SWATH-MS in combination with the neuronal culture model system generate a powerful and versatile tool for molecular profiling in a broad spectrum of applications. This chapter describes a protocol for the SWATH-MS quantification of the primary neuronal culture proteome that can be adapted for the in-depth exploration of the molecular mechanisms of interest.

2 Materials

2.1 Sample Preparation

1. PBS: Dulbecco's Phosphate Buffered Saline 1× (no calcium, no magnesium, Gibco, Life Technologies).

2. Protease inhibitor cocktail tablet (Complete EDTA free, Roche).

3. Loading buffer 1× [12]: 0.05 M Tris–HCl pH 6.8, 2% SDS, 10% glycerol, 0.1 M DTT, 0.001% bromophenol.

4. 30% Acrylamide/Bis Solution 37.5:1 (Bio-Rad).

5. 2,2,2-Trichloroethanol (TCE, Sigma-Aldrich).

6. Gel Doc EZ Imager (Bio-Rad).

2.2 SDS-PAGE and Trypsin Digestion

1. Vertical electrophoresis system (e.g., Mini-PROTEAN Tetra Electrophoresis System from Bio-rad).

2. Fixation solution: 50% ethanol/3% phosphoric acid (v/v) (BioUltra >85%, Sigma-Aldrich).

3. Coomassie Brilliant Blue G (Sigma-Aldrich).

4. MultiScreen-HV 96-well filter-plate (0.45 μm, Millipore) and 96 DeepWell plates (Eppendorf).

5. 50 mM ammonium bicarbonate pH 7.8–8.0 (99% NH_4HCO_3, Fluka).

6. 50 mM NH_4HCO_3/50% (v/v) acetonitrile (LC-MS Ultra CHROMASOLV, Fluka).

7. 0.1% Trifluoroacetic acid/50% acetonitrile (TFA, Sigma-Aldrich) and 0.1% TFA/80% acetonitrile.

8. Trypsin/Lys-C Mix solution 6.7 mg/mL (Mass Spec Grade, Promega) in 50 mM ammonium bicarbonate.

2.3 LC-MS Analysis

1. HPLC solvent A: 2% acetonitrile, 0.1% formic acid (Sigma-Aldrich).

2. HPLC solvent B: 99.9% acetonitrile, 0.1% formic acid.

3. HRM calibration kit (Biognosys).

4. Ultimate 3000 LC system (Thermo Scientific).

5. 5 mm C18 PepMap 100 trapping column (300 μm i.d., 5 μm particle size, 100 Å, Thermo Scientific).

6. Homemade 200 mm C18 Alltima analytical column (100 μm i.d., 3 μm particle size).

7. TripleTOF 5600 mass spectrometer (SCIEX) with integrated DuoSpray ion source (micro-spray needle).

8. 500 μL conical vials (Fisher Scientific).

3 Methods

3.1 Sample Preparation

1. Mouse hippocampal primary neurons were cultured at embryonic day 18 [13]. Cells were plated at a seeding density of 300 k/well in 12-well plates coated with poly-d-lysine (Sigma-Aldrich) and 5% heat-inactivated horse serum (Invitrogen). Neurons were incubated at 37 °C 5% CO_2 until

the developmental stage of interest (e.g. 14 days for mature neurons).

2. At the day of the cell collection, prepare ice-cold PBS 1× with and without protease inhibitor (1 mL each per well) and place the culture plate on ice (*see* **Note 1**).

3. Remove the culture medium and wash each well two times with 500 μL ice-cold PBS in order to remove proteins present in the medium.

4. Add 500 μL of PBS with protease inhibitor and carefully de-attach the neurons from the bottom of the wells with the help of a scraper (*see* **Note 2**).

5. Transfer the cells to 1.5 mL tubes and rinse again each well with 500 μL of PBS with protease inhibitor to collect the cells remaining.

6. Centrifuge the samples for 5 min at 3000 × g at 4 °C to pellet the cells.

7. Remove the supernatant and resuspend the cells in 15 μL of SDS sample-loading buffer. Pipette several times to completely lyse the neurons and heat these at 90 °C for 5 min.

8. To block cysteine residues, add 3 μL of 30% acrylamide and incubate at room temperature for 30 min [14].

9. To ensure equal amounts of protein over all the samples, the protein concentration of each sample can be normalized by running a SDS polyacrylamide gel prior to the trypsin diges-tion. Prepare a 10% SDS polyacrylamide gel containing 0.5% trichloroethanol (TCE) (*see* **Note 3**). Run 1 μL per sample in the gel at 100 V for 45 min. Scan the gel in a Gel Doc EZ Imager (Bio-Rad). The images can be analyzed with Image Lab software to compare and correct the total protein amount between samples.

3.2 In-Gel Digestion

1. Run the proteins very briefly at 100 V for 10–20 min until the bromophenol front reach about 1 cm down the 10% SDS poly-acrylamide gel.

2. Remove the gel and immerse it in the fixation solution overnight.

3. Pour the fixation solution off, wash the gel twice for 15 min each in excess water.

4. Stain briefly with colloidal Coomassie Brilliant Blue G for a few minutes till faint protein bands appear.

5. Cut each sample lane into small pieces with a scalpel and trans-fer them to the wells of a MultiScreen-HV 96-well filter-plate.

6. Place a 96 DeepWell plate below the filter plate to collect the waste solution.

7. Add 150 μL 50% acetonitrile in 50 mM ammonium bicarbonate to each well and incubate for 2 h to destain the gel pieces, centrifuge the plate at 200 × g for 1 min to remove the solution.

8. Add 150 μL 100% acetonitrile to each well, incubate for 10 min, and centrifuge the plate.

9. Add 150 μL 50 mM ammonium bicarbonate to each well to rehydrate the gel pieces in 5 min, centrifuge the plate at 200 × g for 1 min.

10. Repeat the destaining cycle once, or till the gel pieces are totally destained.

11. After dehydration in 150 μL 100% acetonitrile, re-swell the gel pieces with 120 μL Trypsin/Lys-C Mix solution and incubate overnight in a humidified chamber at 37 °C (*see* **Note 4**).

12. Place a clean 96 DeepWell plate below the filter plate, and centrifuge to collect the peptides there.

13. Add 200 μL 50% acetonitrile in 0.1% TFA to each well, incubate for 5 min, and centrifuge.

14. Add 200 μL 80% acetonitrile in 0.1% TFA to each well, incubate for 5 min, and centrifuge.

15. The collected solution in the DeepWell plate contains the digested peptides. Transfer the solution to an Eppendorf tube, dry in a speedvac and store at −20 °C untill SWATH analysis is performed.

16. A spectral library should contain a thorough coverage of peptides generated by DDA analysis, which requires more input material and better peptide fractionation prior to MS analysis. There are two widely used strategies for fractionation:

    (a) The gel-based approach. Allow at least 6 samples to run until the end of the SDS polyacrylamide gel for better separation of proteins according to size. Cut each sample lane into 6–10 slices of equal size, and chop each of these into small pieces. Pool the gel pieces corresponding to the same fraction from different samples into the same well on the 96-well filter plate. Continue the protocol as described above, and run each fraction independently by DDA in the mass spectrometer.

    (b) The high-pH column fractionation approach. The protocol from Chapters 5 and 10 can be followed. Alternatively, the Pierce high-pH reversed-phase peptide fractionation kit can be used to separate the proteolytically digested proteins samples into 8 fractions of peptides and subject them to LC-MS/MS separately.

3.3 LC-MS
Analysis: SWATH

1. Redissolve the peptides in 2% acetonitrile/0.1% formic acid solution containing iRT reference peptides (the HRM calibration kit from Biognosys, *see* **Note 5**). The volume to dissolve peptides should be compatible for the LC loading volume.

2. Transfer the solution to a conical vial, and inject the sample into the Ultimate 3000 LC system.

3. Peptides are trapped on a 5 mm C18 PepMap 100 column for 5 min, and separated on a homemade 200 mm C18 Alltima column. Perform the reversed-phase liquid chromatography by linearly increasing the acetonitrile concentration in the mobile phase at a flow rate of 5 μL/min; from 5% to 22% in 88 min, to 25% at 98 min, to 40% at 108 min, and to 95% in 2 min.

4. The separated peptides are electro-sprayed into the TripleTOF 5600 MS (SCIEX) with a micro-spray needle (at 5500 V).

5. The mass spectrometer is set in data-independent acquisition at high sensitivity and positive mode under the following parameters (optimized for the TripleTOF 5600 and type of samples, Fig. 1):

    (a) Parent ion scan of 100 ms (mass range of 350–1250 Da).

    (b) SWATH window of 8 Da, giving a reasonable number of fragment ions per window, with an overlap of 1 Da between consecutive windows (*see* **Note 6**).

    (c) Use SWATH mass range between 450 and 770 *m/z*, which covers the most part of the peptides (*see* **Note 7**).

    (d) MS/MS scan time of 80 ms per window (range 200–1800 Da), to keep the cycle time low without compromising the sensitivity. A reduced cycle time results in more data points per eluting peptide (measurements in retention time dimension), which improves the accuracy of identification and quantification downstream data analysis.

    (e) The collision energy for each window is determined for a 2+ ion centered upon the window, with a spread of 15 eV.

6. The data can be analyzed using Specter [16] or Spectronaut [10] from Biognosys. For the latter, use default parameters with the following considerations:

    (a) To correct the differences in LC retention time, select dynamic retention time prediction to allow nonlinear alignment of precursor retention times between the spectral library and the SWATH data by segmented regression [6, 10].

    (b) To correct differences in the amount of input sample loaded, enable cross-run normalization.

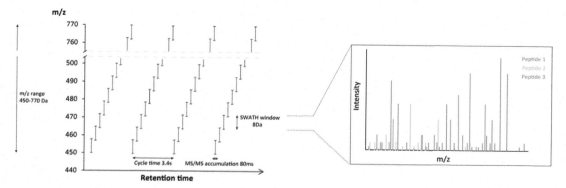

Fig. 1 SWATH MS data-independent acquisition strategy with example parameters. The total *m/z* range (450–770 Da) is subdivided in consecutive and overlapping SWATH windows of 8 Da. All the peptides contained in each window will be simultaneously fragmented and analyzed by MS/MS scan (80 ms). This cycle (3.4 s) will be repeated across the entire chromatogram. The total cycle time is about 3.4 s, which generates around 10 measurement points for the average peptide with elution time of 30 s. In total, the run time is 130 min per sample. Adapted from [15]

(c) To discriminate high confidence peptides, the use of peptides with Q-value ≤ 0.01 in several replicates is recommended for quantification (which represent the peptide identification score FDR).

(d) Analysis results can be exported for further process.

3.4 LC-MS Analysis: Spectral Library

1. To generate the spectral library, the fractionated sample peptides should be dissolved as described above.

2. The conditions of the liquid chromatography have to be the same as for the SWATH samples.

3. In this case, the mass spectrometer must be operated in data-dependent mode. The TripleTOF 5600 acquires a single MS full scan (*m/z* 350–1250, 200 ms) follow by MS/MS of the 20 most abundant precursors (*m/z* 200–1800, 100 ms) at high-sensitivity mode in low resolution (precursor ion gt;100 counts/s, charge state from +2 to +5).

4. CID is performed using rolling collision energy and a spread energy of 15 eV.

5. The exclusion time window is set to 16 s once a peptide is fragmented.

6. Analyze the mass spectrometer output files with MaxQuant software [17] to identify peptides and proteins:

   (a) Search against the Biognosys iRT fasta database and the most updated UniProt mouse proteome, including both canonical and isoform sequences.

   (b) Select propionamide (C) as the fixed modification and Methionine oxidation and N-terminal acetylation as variable modifications (*see* **Note 8**).

(c) Set the minimum peptide length to 6, with at most one miss-cleavage allowed.

(d) For both peptide and protein identification a false discovery rate is set to 1%.

(e) MaxQuant search results can subsequently be imported as DIA spectral library into Spectronaut with default settings.

4   Notes

1. To reduce molecular events associated with cell death as much as possible, prepare all the materials and ice-cold solutions in advance before taking the cells out of the incubator.

2. The purpose of this step is not to lyse the cells, the monolayer of neurons cultured in these conditions will easily detach by gently scraping the bottom of the well.

3. TCE allows the rapid visualization of the protein bands after exposure to UV light [18].

4. Additionally, 50 μL 50 mM ammonium bicarbonate can be added to each sample to avoid the sample dehydration.

5. The iRT kit is a mix of synthetic peptides designed to elute at various points throughout retention time. Spiking these into samples allows for robust normalization of chromatographic differences between mass spectrometric acquisitions, increasing the accuracy of downstream quantitative analysis [19].

6. While the original study successfully reported up to 25 Da windows [15], a more narrow selection provides a deeper protein profiling in highly complex samples [20].

7. The mass range can be selected considering the precursor mass histogram of the input samples (e.g., [5], Sup. Fig. 1). Notice that the miss-match between MS1 and MS2 m/z range barely affect the end results or duty-cycle of the mass spectrometer.

8. Fixed modifications must be selected according to the sample preparation protocol applied (Propionamide in case of the in-gel digestion method described above).

References

1. Dieterich DC, Kreutz MR (2016) Proteomics of the synapse—a quantitative approach to neuronal plasticity. Mol Cell Proteomics 15(2):368–381

2. Pandya NJ, Koopmans F, Slotman JA, Paliukhovich I, Houtsmuller AB, Smit AB, Li KW (2017) Correlation profiling of brain subcellular proteomes reveals co-assembly of synaptic proteins and subcellular distribution. Sci Rep 7(1):1–11

3. Michalski A, Cox J, Mann M (2011) More than 100,000 detectable peptide species elute in single shotgun proteomics runs but the majority is Inaccessible to data-dependent LC–MS/MS. J Proteome Res 10(4):1785–1793

4. Law KP, Lim YP (2013) Recent advances in mass spectrometry: data independent analysis and hyper reaction monitoring. Expert Rev Proteomics 10(6):551–566

5. Koopmans F, Ho JTC, Smit AB, Li KW (2017) Comparative analyses of data independent acquisition mass spectrometric approaches: DIA, WiSIM-DIA and untargeted DIA. Proteomics 18:1700304

6. Bruderer R, Bernhardt OM, Gandhi T, Reiter L (2016) High-precision iRT prediction in the targeted analysis of data-independent acquisition and its impact on identification and quantitation. Proteomics 16(15–16):2246–2256

7. Wang J, Tucholska M, Knight JDR, Lambert J-P, Tate S, Larsen B, Gingras A-C, Bandeira N (2015) MSPLIT-DIA: sensitive peptide identification for data-independent acquisition. Nat Methods 12(12):1106–1108

8. Tsou C-C, Tsai C-F, Teo GC, Chen Y-J, Nesvizhskii AI (2016) Untargeted, spectral library-free analysis of data-independent acquisition proteomics data generated using Orbitrap mass spectrometers. Proteomics 16(15–16):2257–2271

9. Li Y, Zhong C-Q, Xu X, Cai S, Wu X, Zhang Y, Chen J, Shi J, Lin S, Han J (2015) Group-DIA: analyzing multiple data-independent acquisition mass spectrometry data files. Nat Methods 12(12):1105–1106

10. Bruderer R, Bernhardt OM, Gandhi T, Miladinović SM, Cheng L-Y, Messner S, Ehrenberger T, Zanotelli V, Butscheid Y, Escher C, Vitek O, Rinner O, Reiter L (2015) Extending the limits of quantitative proteome profiling with data-independent acquisition and application to acetaminophen-treated three-dimensional liver microtissues. Mol Cell Proteomics 14(5):1400–1410

11. Bruderer R, Sondermann J, Tsou C-C, Barrantes-Freer A, Stadelmann C, Nesvizhskii AI, Schmidt M, Reiter L, Gomez-Varela D (2017) New targeted approaches for the quantification of data-independent acquisition mass spectrometry. Proteomics 17(9):1700021

12. Laemmli UK (1970) Cleavage of structural proteins during the assembly of the head of bacteriophage T4. Nature 227(5259):680–685

13. Gonzalez-Lozano MA, Klemmer P, Gebuis T, Hassan C, van Nierop P, van Kesteren RE, Smit AB, Li KW (2016) Dynamics of the mouse brain cortical synaptic proteome during postnatal brain development. Sci Rep 6(1):35456

14. Mineki R, Taka H, Fujimura T, Kikkawa M, Shindo N, Murayama K (2002) In situ alkylation with acrylamide for identification of cysteinyl residues in proteins during one- and two-dimensional sodium dodecyl sulphate-polyacrylamide gel electrophoresis. Proteomics 2(12):1672–1681

15. Gillet LC, Navarro P, Tate S, Röst H, Selevsek N, Reiter L, Bonner R, Aebersold R (2012) Targeted data extraction of the MS/MS spectra generated by data-independent acquisition: a new concept for consistent and accurate proteome analysis. Mol Cell Proteomics 11(6):O111.016717

16. Peckner R, Myers SA, Jacome ASV, Egertson JD, Abelin JG, MacCoss MJ, Carr SA, Jaffe JD (2018) Specter: linear deconvolution for targeted analysis of data-independent acquisition mass spectrometry proteomics. Nat Methods 15(5):371–378

17. Cox J, Mann M (2008) MaxQuant enables high peptide identification rates, individualized p.p.b.-range mass accuracies and proteome-wide protein quantification. Nat Biotechnol 26(12):1367–1372

18. Ladner CL, Yang J, Turner RJ, Edwards RA (2004) Visible fluorescent detection of proteins in polyacrylamide gels without staining. Anal Biochem 326(1):13–20

19. Escher C, Reiter L, Maclean B, Ossola R, Herzog F, Maccoss MJ, Rinner O (2014) Using iRT, a normalized retention time for more targeted measurement of peptides. Proteomics 12(8):1111–1121

20. Kang Y, Burton L, Lau A, Tate S (2017) SWATH-ID: An instrument method which combines identification and quantification in a single analysis. Proteomics 17(10):1500522

Chapter 12

Quantitative Analysis of Mass Spectrometry-Based Proteomics Data

Thang V. Pham and Connie R. Jimenez

Abstract

This chapter guides the user through an analysis pipeline that includes preprocessing raw mass spectrometry data into a user-friendly quantitative protein report and statistical analysis. We use a publicly available dataset as a working example that covers two prominent strategies for mass spectrometry-based proteomics, the extensively used data-dependent acquisition (DDA) and the emerging data-independent acquisition (DIA) technology. We use MaxQuant for DDA data and Spectronaut for DIA data preprocessing. Both software packages are well-established tools in the field. We perform subsequent analysis in the R software environment which offers a large repertoire of tools for data analysis and visualization. The chapter will aid lab scientists with some familiarity with R to reproducibly analyze their experiments using state-of-the-art bioinformatics methods.

Key words Cluster analysis, Data-dependent acquisition, Data-independent acquisition, Ion-intensity quantification, Limma, Mass spectrometry, MaxQuant, Proteomics, Spectronaut, Statistical analysis

1 Introduction

The liquid chromatography tandem mass spectrometry platform (LC-MS/MS) is a powerful technology for quantitative proteomics [1]. Data-dependent acquisition (DDA) has been a method of choice as it offers robust identification and quantification of thousands of proteins from a single injection of a complex biological sample (*see* Chapters 5, 8, and 14 for protocol). The data-independent acquisition (DIA) approach has recently emerged as a strong alternative thanks to its ability to provide a more complete the data matrix by combining unbiased, broad range precursor ion fragmentation and targeted data extraction [2] (also known as SWATH, *see* Chapter 11 for experimental details). In this chapter, we aim to guide the user through bioinformatics analysis steps typically involved in both DDA and DIA proteomics experiments.

Bioinformatics analysis can be broadly divided into three stages. In the first stage, mass spectrometry signals are mapped to

Ka Wan Li (ed.), *Neuroproteomics*, Neuromethods, vol. 146,
https://doi.org/10.1007/978-1-4939-9662-9_12, © Springer Science+Business Media, LLC, part of Springer Nature 2019

peptide sequences for identification and quantification. For DDA, this involves matching the experimental data against a database of protein sequences and a decoy database of revered sequences to control for false discovery. For DIA, peptides from a spectral library are matched against a set of mixed spectra of multiple peptides. To increase the quality of the matching, synthetic peptides can be spiked in to provide retention time standards for a better calibration of the elution time of the complex peptide mixture. Identified peptides are rolled up to protein-level quantification. There are a number of well-established software packages for data preprocessing. For DDA, the MaxQuant software with an integrated search engine [3] or the IDPicker workflow [4] with multiple peptide search engines [5, 6] can be used. For DIA, Spectronaut [7] and OpenSwath [8] can be considered. In this chapter, we will use MaxQuant and Spectronaut because of their availability and our familiarity with the tools.

In the second stage, the preprocessed data are used for cluster analysis and statistical significance analysis. The data can be viewed in office software tools such as Excel which allows basic visualization and numeric calculation. The Perseus tool [9] is another platform developed in tandem with MaxQuant. Spectronaut has built-in functions for analysis such as performing the t-test. Nevertheless, the R software environment offers a large repository of tools for analysis and visualization. For spectral count data we have developed the beta-binomial test [10, 11]. Software tools developed for RNA-Seq data analysis such as edgeR [12] and DESeq2 [13] can also be employed. For ion-intensity-based quantitation, we have developed a statistical model for technical variation that can be used in the generic s-test [14] to model missing data. Another option is limma [15] with data imputation. For DIA, the R package MSstats [16] can be used. We will demonstrate the use of limma, with and without data imputation.

The third stage of analysis consists of various forms of integration with existing knowledge, either with domain experts or using knowledge databases such as gene ontology and protein-protein interaction networks. We do not cover this stage of analysis. We are aware of R packages for gene ontology analysis, gene set enrichment analysis, and network analysis. The results in R can also be exported to file formats suited for network analysis tools such as Cytoscape [17].

There are two modes of quantitation in DDA for a label-free proteomics workflow, spectral counting and ion-intensity-based quantification. Spectral counting based on the number of MS/MS spectra observed for a protein is robust and immediately available after the database search [18]. Ion-intensity-based quantitation is more elaborate, requiring signal processing at the MS1 or MS2 level. Nevertheless, software packages such as MaxQuant and

IDPicker have made this step a rather simple procedure to the end user. The raw signal intensities at the peptide level are aggregated and normalized to protein level called MaxLFQ [19]. For DIA, ion-based quantitation is used where MS/MS fragment intensities are combined to infer protein-level quantitation. We will use ion-intensity quantitation for our analysis in this chapter with a side note on spectral counting.

As a working example, we have downloaded a publicly available LC-MS/MS dataset that was used in a benchmark experiment for label-free DDA and DIA proteomics [7]. The raw data set for each type of acquisition contains 24 LC-MS/MS runs of 8 biological replicates and 3 technical replicates. For each biological sample, 12 proteins were spiked in at different concentrations. In total, there are 28 possible pairwise comparisons, which can be used to compare and contrast different analysis methods.

2 Materials

1. We use a spike-in dataset in [7] as a working example. The raw LC-MS/MS data are available at the public repository PeptideAtlas (http://www.peptideatlas.org/PASS/PASS00589). On the downloading site, the DDA raw data files are *B_D140314_SGSDSsample$^{(\#)}$_R0$^{(\#\#)}$_MSG_T0.raw* where the sample number $^{(\#)}$ runs from 1 to 8, and the replicate number $^{(\#\#)}$ runs from 1 to 3. Raw DIA data are of the format *B_D140314_SGSDSsample$^{(\#)}$_R0$^{(\#\#)}$_MHRM_T0.raw*.

2. The fasta sequences of the spike-in proteins are in the file *QS-spike-in-proteins.fasta* on the web location above. Fasta sequences of the peptides used for retention time alignment (the iRT peptides) are available from Biognosys (https://www.biognosys.com/). A fasta file of the human proteome is available from Uniprot (http://www.uniprot.org/).

3. The MaxQuant software tool and tutorials are available online (http://www.maxquant.org). There is also an active user group for troubleshooting. We use MaxQuant version 1.6.0.16.

4. Spectronaut is commercially available from Biognosys. We use Spectronaut version 11.0. Consult the Spectronaut user manual and help that comes with the software.

5. The R software environment is freely available from (https://cran.r-project.org/). We use R version 3.4.3. This protocol also uses the limma R package which can be installed from within R.

6. Optionally, we find it convenient to use a spreadsheet program such as Microsoft Excel or OpenOffice Calc to explore data tables, and a text editor such as Notepad. These tools are complementary to the R console-based environment.

3   Methods

3.1   DDA Data Preprocessing Using MaxQuant

The followings are the main steps if one does not wish to deviate considerably from the MaxQuant default values. For a more detailed guide and troubleshooting, consult the MaxQuant user group or a recent protocol paper [20].

1. Copy the raw DDA data to a working folder (*see* **Note 1**).

2. Concatenate the spike-in protein fasta sequences, the iRT peptide fasta sequences, and the Swissprot human proteome fasta sequences into a single fasta file using a text editor or a shell command line.

3. Start MaxQuant and configure the sequence database to include the combined fasta file just created.

4. Load the raw data folder into MaxQuant.

5. Load an experimental design file that specifies which raw files belonging to which biological samples. One might export a so-called experimental design template, use an external spreadsheet editor to fill out the details, and subsequently load the experimental design back into MaxQuant. In our example, we name the 24 samples by the combination of sample from 1 to 8 and replicate from 1 to 3. Since there is no fractionation, all fraction numbers are assigned to 1.

6. Specify the concatenated fasta sequence file for database search.

7. Specify the number of CPU cores available for the analysis and start the processing.

MaxQuant produces several text files in the *combined\txt* subfolder of the raw file folder. We use the *proteinGroups.txt* for protein quantitation.

3.2   DIA Data Processing Using Spectronaut

1. Copy the raw DIA data to a working folder.

2. Generate a spectral library from MaxQuant analysis of DDA data (*see* **Note 2**).

   (a) Import the text output folder of MaxQuant.

   (b) Review the default parameters.

   (c) Enter a name for the library to be created.

   (d) Select the combined fasta file previously created.

   (e) Start the processing.

      Spectronaut creates a spectral library from the MaxQuant search and stores the result in its default directory (*see* **Note 3**).

3. Start a new DIA analysis.

(a) Load the raw DIA data folder.

(b) Enter an experiment name for the analysis.

(c) Select the combined fasta file.

(d) Select the spectral library.

(e) Create an experiment design. Like in MaxQuant, we find it convenient to export a template file, use a spreadsheet tool to edit the template, and import the result back. The important columns are "Replicate", "Condition", and "Label". In a profiling experiment of independent samples, the "Replicate" number is 1 and "Condition" and "Label" can be identical as the raw file names.

(f) Start the analysis.

4. Export a protein report for further analysis in R.

(a) Export a protein report. In this so-called wide format, each row is a protein group and each column either a sample or some annotation for the protein group. For annotation, we select in addition to the protein group IDs, gene names (PG.Genes), protein names (PG.ProteinNames), q-values (PG.Qvalue), and Uniprot identifiers (PG. UniProtIds). We name the export *proteinGroups-dia.txt*.

(b) Save the project file (.sne) for future reference.

Unlike MaxQuant, one must explicitly export the output needed (*see* **Note 4**). Also, it is a good practice to examine the various quality control figures produced by the software in the post-analysis tab.

3.3 Descriptive Data Analysis

3.3.1 Basic Data Handling in R

1. Copy the *proteinGroups.txt* file from the DDA analysis and the *proteinGroups-dia.txt* file from DIA analysis to a working directory. Both files are tab-separated files and can be explored in Excel.

2. Start R and change to the working directory above. One should see the two files in the step above by listing the directory (by entering *dir()* from the R console).

3. Read in the DDA protein report and keep rows not marked as reversed entries (in the MaxQuant output, these are empty cells in the column "Reverse"). Locate columns starting with *LFQ* for quantitative values as follows.

```
dda <- read.delim("./proteinGroups.txt")

dda <- subset(dda, Reverse == "")     # remove reversed entries

rownames(dda) <- dda[, "Protein.IDs"] # use protein group ids as row names

lfq <- grep("^LFQ", colnames(dda))

dda_log2 <- log2(dda[, lfq])

dda_log2[dda_log2 == -Inf] <- NA
```

All functions are elementary R functions with a possible exception of *grep()*. This function uses the so-called regular expression to search for column names starting with *LFQ*, and stores the matched indexes in the variable *lfq*. Type *lfq* in the R console to see its value. Thus, *dda[, lfq]* contains the subset of the whole table that contains the MaxLFQ values of the samples. Finally, the log2-transformed values are stored in *dda_log2*. Check the dimension of the data table by *dim(dda_log2)*. (For spectral counting quantification, *see* **Note 5**).

4. Check the expression of the spike-in proteins.

```
spike_ins <- c("P02754", "P80025", "P00921", "P00366", "P02662", # mix 1

        "P61823", "P02789", "P12799", "P02676", "P02672", # mix 2

        "P02666", "P68082")                    # mix 3

# check the spike-ins

for (i in 1:length(spike_ins)) {

    cat(grep(spike_ins[i], rownames(dda_log2)), "\n")

}

# change row names to the spike-ins names

for (i in 1: length(spike_ins)) {

    rownames(dda_log2)[grep(spike_ins[i], rownames(dda_log2))] <- spike_ins[i]

}

dda_log2[spike_ins,]
```

We look for the spike-in proteins in the row names of the data table, i.e. the protein group identifiers. In the protein assembly step, it is possible that a protein is present in multiple protein groups. Nevertheless, it is not the case in the current analysis. MaxQuant detects all 12 spike-in proteins. Each spike-in protein is present in one and only one entry. We rename the row names of protein groups containing the spike-in to the spike-in proteins. Hence, the entry "CON__ P02754;P02754" become "P02754". (This is because a spike-in protein is also marked as contaminant in the MaxQuant search).

5. Read in the DIA protein report. Note that in the Spectronaut analysis, next to the protein groups IDs ("PG. ProteinAccessions"), we chose four other annotation columns. Thus, the first 5 columns are annotation and the remaining 24 columns (from 6 to 29) are expression values.

```
dia <- read.delim("proteinGroups-dia.txt", na.strings = "Filtered")

rownames(dia) <- dia[, "PG.ProteinAccessions"]

dia_log2 <- log2(dia[, 6:29])

dia_log2[dia_log2 == -Inf] <- NA

# check spike-ins

for (i in 1: length(spike_ins)) {

   cat(grep(spike_ins[i], rownames(dia_log2)), "\n")

}

# change row names to the spike-ins names

for (i in 1: length(spike_ins)) {

   rownames(dia_log2)[grep(spike_ins[i], rownames(dia_log2))] <- spike_ins[i]

}

dia_log2[spike_ins,]
```

The DIA data handling is similar to that for DDA data. Note that missing values are denoted as "Filtered" in the original data file.

3.3.2 Missing Values in DDA and DIA Data

sum(is.na(dda_log2))

sum(is.na(dia_log2))

heatmap(**as.matrix**(dda_log2), Rowv = NA, Colv = NA, scale = "none", labRow = NA,

labCol = **paste0**("s", **rep**(1:8, each = 3), "_r", **rep**(1:3, 8)), col = "gray")

heatmap(**as.matrix**(dia_log2), Rowv = NA, Colv = NA, scale = "none", labRow = NA,

labCol = **paste0**("s", **rep**(1:8, each = 3), "_r", **rep**(1:3, 8)), col = "gray")

The first two statements count the number of missing values in DDA and DIA data. One observes that the number of missing values in DDA is much higher than that of DIA data. Figure 1 illustrates the difference in two heatmaps where detected proteins are shown in gray and missing values are white spaces. By default, the *heatmap()* function produces clusterings of rows and columns. We disable this feature by assigning an NA value to both parameters *Rowv* and *Colv*.

(A)   **(B)**

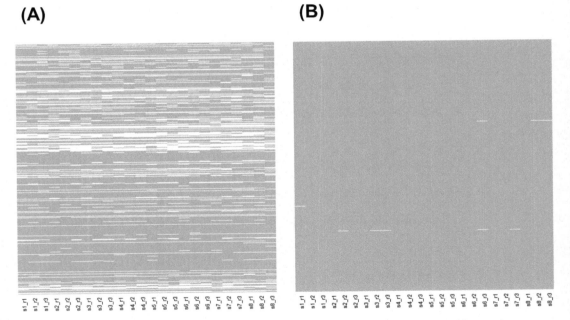

Fig. 1 Missing values patterns, (**a**) DDA data and (**b**) DIA data. Missing values are denoted by empty spaces

3.3.3   Visualization of Spike-In Proteins

```
# spike-ins in DDA

par(mfrow = c(3,1),  mar = c(1, 6, 1, 1))

matplot(t(dda_log2[spike_ins[1:5],]), type = c("b"), pch = 1, col = 1:5,
    ylab = "MIX1 - log2 LFQ", xaxt="n")

legend("topleft", legend = spike_ins[1:5], col = 1:5, pch = 1)

matplot(t(dda_log2[spike_ins[6:10],]), type = c("b"), pch = 1, col = 6:10,
    ylab = "MIX2 - log2 LFQ", xaxt="n")

legend("bottomleft", legend = spike_ins[6:10], col = 6:10, pch = 1)

par(mar = c(4, 6, 1, 1)) # more space for the x axis of the bottom panel

matplot(t(dda_log2[spike_ins[11:12],]), type = c("b"), pch = 1, col = 11:12,
    ylab = "MIX3 - log2 LFQ", xaxt="n")

axis(side = 1, at = (0:7)*3 + 2, labels = paste0("s", 1:8))

legend("topleft", legend = spike_ins[11:12], col = 11:12, pch = 1)


# spike-ins in DIA

par(mfrow = c(3,1), mar = c(1, 6, 1, 1))

matplot(t(dia_log2[spike_ins[1:5],]), type = c("b"), pch = 1,col = 1:5,
    ylab = "MIX1 - log2 Intensity", xaxt = "n")

legend("topleft", legend = spike_ins[1:5], col = 1:5, pch = 1)

matplot(t(dia_log2[spike_ins[6:10],]), type = c("b"), pch = 1, col = 6:10,
    ylab = "MIX2 - log2 Intensity", xaxt = "n")

legend("bottomleft", legend = spike_ins[6:10], col = 6:10, pch = 1)

par(mar = c(4, 6, 1, 1)) # more space for the x axis of the bottom panel

matplot(t(dia_log2[spike_ins[11:12],]), type = c("b"), pch = 1, col = 11:12,
    ylab = "MIX3 - log2 Intensity", xaxt = "n")

axis(side=1, at=(0:7)*3 + 2, labels = paste0("s", 1:8))

legend("topleft", legend = spike_ins[11:12], col = 11:12, pch = 1)
```

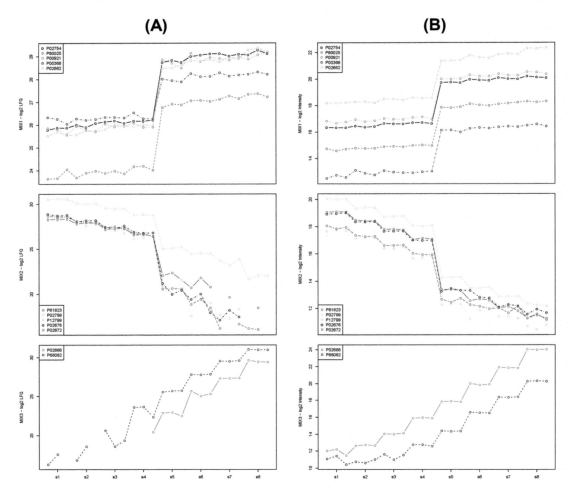

Fig. 2 Plots of spike-in proteins in three different mixes, (**a**) DDA data and (**b**) DIA data. All 12 markers are recovered in all samples in the DIA mode

Figure 2 shows the result of the script. The main function is *mat-plot()* that plots rows of a data matrix. We split the plotting area in to three parts, corresponding to the three mixes of the spike-ins.

Clustering is another powerful way to visualize the data. In the following, we show the heatmap and hierarchical clusterings using the basic *heatmap()* function. Alternatively, the user might want to explore other visualization packages such as gplots [21] for this purpose.

```
heatmap(as.matrix(dda_log2[spike_ins,]), scale = "row", labRow = spike_ins,

    labCol = paste0("s", rep(1:8, each = 3), "_r", rep(1:3, 8)))

heatmap(as.matrix(dia_log2[spike_ins,]), scale = "row", labRow = spike_ins,

    labCol = paste0("s", rep(1:8, each = 3), "_r", rep(1:3, 8)))
```

Here we scale the data matrix by row, meaning that each protein will be transformed to the so-called z-score values, exhibiting the up and down patterns rather than the expression values. For hierarchical clustering, one can choose different distance functions and linkages. The result of the script is shown in Fig. 3.

3.4 Differential Analysis

3.4.1 Analysis with Limma

We will use the R package limma for analysis. One has to install the package if it is not already in the system (thus, the following statements need to be executed once only, also *see* **Note 6**).

```
source("https://bioconductor.org/biocLite.R")

biocLite("Biobase")

biocLite("limma")
```

First, we compare the three replicates of sample 1 to the three replicates of samples 2 for DIA data. Then, we will analyze DDA data with imputation because of the substantial amount of missing data.

```
require(limma)

require(Biobase)

myeset <- ExpressionSet(assayData = as.matrix(dia_log2[, 1:6]))

groups <- c("group1", "group1", "group1", "group2", "group2", "group2")

design <- model.matrix(~ 0 + groups)

colnames(design) <- c("group1", "group2")

contrast.matrix <- makeContrasts("group2-group1", levels = design)

fit <- lmFit(myeset, design)

fit2 <- contrasts.fit(fit, contrast.matrix)

fit2 <- eBayes(fit2)

limma_output <- topTable(fit2, sort="none", n = Inf)

limma_output[spike_ins,]
```

We create a so-called expression set *myeset* from the data table and specify that the first three columns are labeled as group 1 and the next three columns as group 2. The subsequent statements are limma and R specifics to perform an independent sample test. One can adapt this part for another experimental design such as a paired sample test. The final output is stored in *limma_output*. The p-values are in column "P.Value" and the log2 fold change is in

(A)   **(B)**

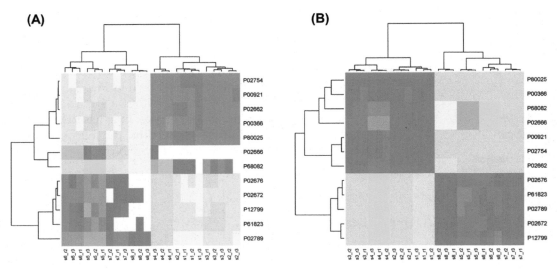

Fig. 3 Heatmap and hierarchical clustering using spike-in proteins, (**a**) DDA data and (**b**) DIA data

"logFC". One can write the output table to a text file to further explore in Excel as follows

write.table(cbind(**dia_log2, limma_output**), **"limma_output.txt"**,

        sep = "\t", row.names = FALSE)

Here we combine the log2 data table and the output by the *cbind()* function.

3.4.2 DDA with Imputation

Data imputation is useful for datasets with a small number of samples. For instance, without imputation there are no *p*-values for black-and-white regulated proteins, and therefore one cannot easily sort the data to prioritize proteins. Constant imputation is simple in R. One can replace all missing values with a small value such as 0. Another option is the half-min imputation which uses a constant value equal to half of the minimum intensity detected in the dataset.

dda_imputed <- dda_log2

dda_imputed[is.na(**dda_log2**)] **<- 0** *# a small value, here 0 is used*

dda_imputed[is.na(**dda_log2**)] **<- 0.5** * min(**dda_log2, na.rm = TRUE**) *# half-min*

A more elaborate method is to draw random values from a normal distribution. In the following we use imputation from a normal distribution with the mean equal the minimum intensity of the data and the standard deviation equal the average of all standard deviations.

```
dda_imputed <- dda_log2

set.seed(1)

global_mean <- min(dda_imputed, na.rm=TRUE)

global_sd <- mean(apply(dda_imputed, 1, sd, na.rm = TRUE), na.rm = TRUE)

number_na <- sum(is.na(dda_imputed))

dda_imputed[is.na(dda_imputed)] <- rnorm(number_na,

              mean = global_mean,

              sd = global_sd)
```

Here the R function *rnorm ()* is used to generate random numbers from a normal distribution. It is essential to set a random seed by the function *set.seed ()* to ensure reproducibility of the analysis. The imputed data *dda_imputed* can be used to create *myeset* as in limma analysis for DIA data.

4   Notes

1. Make sure one has enough space for data preprocessing. MaxQuant requires from two to three times the raw data volume. By default, all temporary files are kept within the raw file folder.

2. Spectronaut keeps intermediate files in temporary directories, by default in the system drive, which is not abundant in space in many systems. It is recommended that the users carefully check this global setting.

3. When contaminants are included in the MaxQuant search, the spectral library creation in Spectronaut might warn that there are entries in the library that are not in the provided fasta file. This warning can be ignored.

4. For MSstats analysis, export the Spectronaut result in the MSstats format.

5. For spectral count analysis, the columns with names starting with "MS/MS count" can be used. Special characters and whitespaces in the column names are automatically converted into dots in R.

6. If limma or some of its required packages are already present in the system, the installation might ask for an update of outdated packages.

References

1. Aebersold R, Mann M (2003) Mass spectrometry-based proteomics. Nature 422:198

2. Gillet LC, Navarro P, Tate S, Röst H, Selevsek N, Reiter L, Bonner R, Aebersold R (2012) Targeted data extraction of the MS/MS spectra generated by data-independent acquisition: a new concept for consistent and accurate proteome analysis. Mol Cell Proteomics 11:O111–016717

3. Cox J, Mann M (2008) MaxQuant enables high peptide identification rates, individualized ppb-range mass accuracies and proteome-wide protein quantification. Nat Biotechnol 26:1367

4. Ma Z-Q, Dasari S, Chambers MC, Litton MD, Sobecki SM, Zimmerman LJ, Halvey PJ, Schilling B, Drake PM, Gibson BW, others (2009) IDPicker 2.0: improved protein assembly with high discrimination peptide identification filtering. J Proteome Res 8:3872–3881

5. Tabb DL, Fernando CG, Chambers MC (2007) MyriMatch: highly accurate tandem mass spectral peptide identification by multivariate hypergeometric analysis. J Proteome Res 6:654–661

6. Kim S, Pevzner PA (2014) MS-GF+ makes progress towards a universal database search tool for proteomics. Nat Commun 5:5277

7. Bruderer R, Bernhardt OM, Gandhi T, Miladinović SM, Cheng L-Y, Messner S, Ehrenberger T, Zanotelli V, Butscheid Y, Escher C, others (2015) Extending the limits of quantitative proteome profiling with data-independent acquisition and application to acetaminophen-treated three-dimensional liver microtissues. Mol Cell Proteomics 14:1400–1410

8. Röst HL, Rosenberger G, Navarro P, Gillet L, Miladinović SM, Schubert OT, Wolski W, Collins BC, Malmström J, Malmström L, Aebersold R (2014) OpenSWATH enables automated, targeted analysis of data-independent acquisition MS data. Nat Biotechnol 32:219

9. Tyanova S, Temu T, Sinitcyn P, Carlson A, Hein MY, Geiger T, Mann M, Cox J (2016) The Perseus computational platform for comprehensive analysis of (prote) omics data. Nat Methods 13:731

10. Pham TV, Piersma SR, Warmoes M, Jimenez CR (2009) On the beta-binomial model for analysis of spectral count data in label-free tandem mass spectrometry-based proteomics. Bioinformatics 26:363–369

11. Pham TV, Jimenez CR (2012) An accurate paired sample test for count data. Bioinformatics 28:i596–i602

12. Robinson MD, McCarthy DJ, Smyth GK (2010) EdgeR: a bioconductor package for differential expression analysis of digital gene expression data. Bioinformatics 26:139–140

13. Love MI, Huber W, Anders S (2014) Moderated estimation of fold change and dispersion for RNA-seq data with DESeq2. Genome Biol 15:550

14. Pham TV, Jimenez CR (2016) Simulated linear test applied to quantitative proteomics. Bioinformatics 32:i702–i709

15. Smyth GK (2004) Linear models and empirical Bayes methods for assessing differential expression in microarray experiments. Stat Appl Genet Mol Biol 3:1–25

16. Choi M, Chang C-Y, Clough T, Broudy D, Killeen T, MacLean B, Vitek O (2014) MSstats: an R package for statistical analysis of quantitative mass spectrometry-based proteomic experiments. Bioinformatics 30:2524–2526

17. Shannon P, Markiel A, Ozier O, Baliga NS, Wang JT, Ramage D, Amin N, Schwikowski B, Ideker T (2003) Cytoscape: a software environment for integrated models of biomolecular interaction networks. Genome Res 13:2498–2504

18. Liu H, Sadygov RG, Yates JR (2004) A model for random sampling and estimation of relative protein abundance in shotgun proteomics. Anal Chem 76:4193–4201

19. Cox J, Hein MY, Luber CA, Paron I, Nagaraj N, Mann M (2014) Accurate proteome-wide label-free quantification by delayed normalization and maximal peptide ratio extraction, termed MaxLFQ. Mol Cell Proteomics 13:2513–2526

20. Tyanova S, Temu T, Cox J (2016) The MaxQuant computational platform for mass spectrometry-based shotgun proteomics. Nat Protoc 11:2301

21. Warnes GR, Bolker B, Bonebakker L, Gentleman R, Huber W, Liaw A, Lumley T, Maechler M, Magnusson A, Moeller S, Bill V (2009) Gplots: Various R programming tools for plotting data. R package version 2.1

## Chapter 13

2D-DIGE as a Tool in Neuroproteomics

Florian Weiland

Abstract

Neuroproteomics encompasses the study of all protein-related dynamics of the nervous system, not only on a morphological level in development and disease but also functional aspects of the process of learning as well as changes in clinical conditions like depression and addiction. Detection of these changes in protein abundance under defined conditions is the field of differential proteome analysis, with two-dimensional difference gel electrophoresis (2D-DIGE) currently being the most comprehensive technique to quantify these changes while retaining information about isoforms and posttranslational modifications. This chapter focusses on technical aspects of 2D-DIGE as well as a brief overview of successful applications of this technique in neuroproteomics.

Key words 2D-DIGE, Proteomics

1 Introduction

2D-DIGE is an established tool for relative protein quantification and was introduced by Ünlü et al. in 1997. The technique itself encompasses the labeling of protein samples with fluorescent tags and simultaneous separation of two samples on a single 2D gel [1]. This approach was extended by Alban et al. to incorporate an internal pooled standard (IPS), allowing for accurate and reproducible quantification of proteins over multiple gels [2]. 2D-DIGE is a top-down proteomics method, which means that the proteins remain intact during analysis. This enables quantification of changes in posttranslational modifications, truncation and degradation events, splice variants as well as changes in protein isoelectric point (pI) due to variations/mutations in the genetic code [3]. This ability to quantify proteoforms is a major advantage in comparison to bottom-up proteomics methods (e.g. LC-MS/MS), where essentially peptides are quantified and, for the majority of proteins, their specific isoform information is lost [4]. The importance of isoform specific quantification is evident by the example of the survivin protein. The wild-type and splice variant ΔEx3 of sur-

Ka Wan Li (ed.), *Neuroproteomics*, Neuromethods, vol. 146,
https://doi.org/10.1007/978-1-4939-9662-9_13, © Springer Science+Business Media, LLC, part of Springer Nature 2019

vivin inhibit apoptotic processes, whereas the 2B isoform confers pro-apoptotic processes [5]. A failure to quantify the correct proteoform can therefore lead to wrong conclusions about activation of cellular pathways and overall reaction of cells to stimuli. Although 2D-DIGE is the most comprehensive method for top-down protein quantification [6], it has limitations in terms of proteins with extreme pIs as well as strongly hydrophobic proteins [7].

2D-DIGE is commonly carried out using either of two experimental approaches: (1) minimal labeling and (2) saturation labeling. Minimal labeling encompasses the derivatization of the ε-amino group of lysine with a fluorescent dye and in term does not alter the protein pI [1]. This approach labels 1–2% of all available lysines, with only one dye attached per protein [1, 8]. Saturation labeling employs fluorescent dyes which labels all free SH groups, rendering this approach an order of magnitude more sensitive than minimal labeling [9] and is commonly used to detect changes in the proteome where only minute sample amounts are available [10, 11] . However, for saturation labeling, the amount of dye, reducing agent, and protein need to be established and optimized beforehand to ensure complete labeling, and is therefore not suited for routine analysis.

2DE is commonly carried out in buffer systems which denature proteins, but is not restricted to this. Experimental setup variations allowing the analysis of native proteins [12–14] and their differential quantification have been described [15], although differential protein complex quantification using fluorescent dyes has so far only been reported when employing a blue-native and subsequent SDS-PAGE dimension [16–19], as the Coomassie blue stain used as charge carrier quenches the fluorescence of the dyes.

This chapter emphasises on sample preparation, 2D gel electrophoresis, image analysis, and a brief look at statistical procedures as well as the application of DIGE in Neuroproteomics.

2   Materials and Methods

2.1   Sample Preparation

The proteome is the set of proteins expressed by a cell under defined conditions and is highly dynamic. The most important aspects of sample preparation, therefore, are to preserve the state of the proteome, render it accessible for analysis, and ensure the extracted proteome actually resembles the (unknown) proteome of interest. Furthermore, a robust sample preparation protocol encompasses the minimally necessary steps to keep technical variance low to avoid masking the biological variance between the proteomes of interest, as well as checkpoints to ensure sample integrity.

The highly dynamic and rapid nature of changes in posttranslational protein modifications, especially phosphorylations [20,

21], exemplify the need for rapid inactivation of all cellular enzymes during sample storage and preparation. Although protease inhibitors are commonly incorporated during cell lysis, their timely inhibition of proteases is questionable [22]. The inhibition of enzymes responsible for phosphorylations, methylations, ubiquitylation, etc., with their respective inhibitors may therefore be problematic as well. Rapid denaturation and/or precipitation to stabilize the proteome has been described in the literature: Prevention of protein degradation by TCA/Acetone precipitation [22] and heat stabilization [23] can be achieved. Stabilization of posttranslational modifications like phosphorylations can be accomplished using trichloroacetic acid & ethanol precipitation [24] or heat stabilization [25], which also preserves the sumoylation status [23].

Sample preparation for differential proteome analysis using 2D-DIGE has special requirements, mainly due to the demands of isoelectric focusing (IEF). Removal of substances incompatible with IEF such as (1) DNA, (2) lipids, and (3) salts needs to be ensured. High concentrations of DNA increase the viscosity of the sample and can lead to the clogging of the polyacrylamide pores of the IPG strip. DNA can be enzymatically removed using nucleases; however, the usage of DNase I is not recommended as this can lead to artifact generation [26]. Alternatively, ultrasound can be applied to shear DNA. High amounts of lipids can "deplete" detergents used in protein solubilisation buffers and can be removed by TCA/acetone precipitation. Further, salt ions need to be removed from the sample, especially when derived from strong acids (e.g. NaCl [27]) (*see* below). This can be easily achieved by TCA/acetone precipitation and sufficient salt depletion should be controlled using a conductivity meter. After removal of interfering substances, proteins for 2D-DIGE are typically solubilized in a 7 M Urea, 2 M Thiourea, 4%(w/v) CHAPS, 30 mM Tris buffer supplemented with 1% phosphatase- and protease inhibitor cocktails, the buffer pH dependents on the type of labeling to be applied (8.5 for minimal-, 6.5–7.5 for saturation labeling experiments). Deionization of the protein solubilization buffer during preparation is recommended to scavenge ammonium cyanate (derived from Urea), preventing protein carbamylation [28]. Buffer should be aliquoted, stored at −80 °C, and should not be reused. Care should be taken when choosing protease inhibitors, as, for example, 4-(2-aminoethyl) benzenesulfonyl fluoride hydrochloride causes protein and charge modification during labeling experiments [29] and should therefore be avoided. Integrity of all samples must be checked before the actual 2D-DIGE experiment. Reproducibility of protein pattern to control for degradation can be established using SDS-PAGE. Resemblance of the proteome to the desired in vivo state should be checked using antibodies against known differential abundant proteins under the applied experimental conditions (if known beforehand).

2.2 Minimal Labeling

Minimal labeling encompasses the derivatization of lysine residues with a fluorescent dye and offers a limit of detection of less than 1 fmol per protein spot with a dynamic range of more than 4 orders of magnitude [8, 30]. The labeling reaction is catalysed by an NHS-ester group, therefore reagents exhibiting primary amine groups need to be avoided or used in a minimal concentration as they compete for the lysine residues (e.g. Tris should be maximal at a concentration of 30 mM). The optimal pH for this type of reaction is 8.5, with a dye to protein ratio of 200 pmol dye to 50 μg protein [8]. This protocol ensures that only 1–2% of all available lysines are labeled, with only one lysine per protein [1, 8]. Overlabeling needs to be avoided, as this will lead to protein precipitation due to the hydrophobic nature of the dyes [1, 31]. The available dyes for minimal labeling exhibit preference for certain proteins [8, 32, 33], therefore either a dye-swap design or the application of a two dye setup instead of a three-dye setup was suggested [32, 34]. The three-dye setup encompasses the co-separation of two samples with an IPS on one 2D gel, while the two dye setup separates one sample together with an IPS (*see* Fig. 1). However, it should be emphasised that a two-dye setup is recommended as a three-dye setup skews the p-value distribution towards 1, with the consequence that q-values to control for the false discovery rate cannot be applied anymore [35]. The dyes are solubilized in anhydrous dimethylformamide (DMF), which needs to be freshly opened as water contaminations will lead to DMF degradation [36]. Before commencing the labeling reaction, the volume of all samples should be normalized as this will cause similar signal intensities of all gels when scanning (own observation). The labeling reaction itself is carried out for 30 min on ice and quenched using lysine. However, there is evidence that the recommended lysine concentration of 10 mM [8] is not sufficient to completely inhibit dye reactivity, leading to cross-labeling artifacts [37, 38]. A 100-fold increased quencher concentration has been recommended [38]; however, this might cause problems with subsequent IEF due to the increased ion concentration. Our own observation is that the addition of reducing agent and 45 min incubation time on ice with subsequent addition of carrier ampholytes before combining the samples quenches the reaction sufficiently.

2.3 Saturation Labeling

Saturation labeling is the technique of choice if the sample to be analyzed is scarce or an increased sensitivity in terms of number of protein spots detectable is necessary. The labeling reaction occurs at the thiol groups of proteins, and its application for proteomics was firstly described by Shaw et al. [9]. By derivatization of all free SH groups via a maleimide reaction, saturation labeling enhances the sensitivity of protein detection by an order of magnitude as compared to minimal labeling [9]. Sulfur bridges are commonly reduced prior to the labeling reaction to be able to access all cyste-

A   Three-dye experimental design

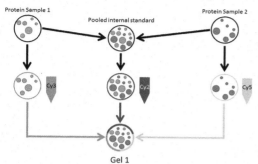

B   Two-dye experimental design

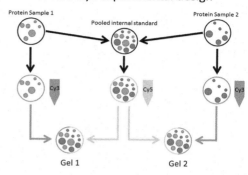

Fig. 1 (**a**) Experimental design incorporating three fluorescent dyes, co-separation of two samples, and pooled internal standard on one 2DE gel. (**b**) Experimental design incorporating two fluorescent dyes, co-separation of one sample, and pooled internal standard on one 2DE gel. From: Arentz, G.; Weiland, F.; Oehler, M. K.; Hoffmann, P., State of the art of 2D DIGE. *Proteomics Clin Appl* 2015, 9, (3-4), 277-88

ines, although omitting this step enables studying of in vivo cysteine modifications [39–41]. Thorough optimization of the saturation labeling process needs to be conducted before attempting to quantify proteins in a large-scale experiment, as a complete labeling of all available cysteine residues is mandatory to avoid artifact spots on the 2D gels. For in-depth guidelines on the optimization of saturation labeling DIGE see the publication of McNamara et al. [42]. Protein identification of protein spots from saturation labeling gels can be achieved either by a preparative gel or, in our observation, by the pooling of corresponding spots from several gels (number depending on spot intensity).

2.4 First Dimension: Isoelectric Focusing

The usage of immobilines (a weak acid or a base) to form pH gradients copolymerized into a polyacrylamide matrix (immobilized pH gradients (IPG) strips) were introduced in 1982 by Bjellqvist et al. [43]. In brief, IEF separates proteins in an electric field depending on their net charge state at a certain pH until the proteins reach a pH where their net charge is zero (pI). Before commencing IEF, IPG strips are rehydrated in a buffer system compatible with high electric fields which additionally keeps proteins denatured and SH-groups reduced or blocked. For these reasons, a 6 M urea, 2 M thiourea, 1% (w/v) CHAPS, 0.5% (v/v) carrier ampholytes, and 200 mM Bis(2-hydroxyethyl) disulphide (HED) buffer system is commonly used. Urea and thiourea facilitate protein unfolding and denaturation, whereas CHAPS is a zwitterionic detergent that keeps proteins solubilized without masking the pI of proteins with excess charge. Carrier ampholytes support the buffering of the pH gradient of the IPG strip and further prevent protein transfer issues into the second dimension caused by hydrophobic, alkaline immobilines [44]. HED prevents re-oxidation of reduced SH-groups [45] as prior reduction/alkylation of proteins using DTT/iodoacetamide before IEF is not recommended as this can cause charge artifacts [46]. After rehydration, which usually takes overnight at room-temperature, the IPG strips need to be briefly immersed in water to remove attaching oil which was used to cover the strip while the rehydration step. An additional gentle blotting step using a moist filter paper to dissolve potential Urea crystals is recommended [47].

Samples are commonly applied using a cup placed on the anodal end of the IPG strip. Here it is important to start the IEF with low electric field strengths, as otherwise salt contaminations can lead to a strong pH drop in the sample application cup and the loss of proteins [27]. A typical protocol for a 24 cm IPG strip with a pH-gradient of 3–11 would encompass the following steps: 1 h at 150 V, 1 h at 300 V, 2.5 h at 600 V, gradient to 10,000 V within 2.5 h, 27,000 Vh at 10,000 V with the current limited in all steps to 50 µA. Paper wicks which are placed between the contact electrodes and IPG strip should to be changed every hour until the voltage gradient phase for superior results [47]. In case of DIGE, IEF needs to be commenced in the dark to prevent dye bleaching. When IEF is finished, the strips can be stored at −80 °C for up to a week, but care must be taken as they turn brittle when frozen.

2.5 Second Dimension: SDS-PAGE

Proteins separated by IEF need to be prepared for SDS-PAGE by immersing the IPG strip in an equilibration solution. Equilibration serves to incorporate proteins in SDS micelles, which enables their transfer from the IPG strip onto the SDS-PAGE gel by an electric field. Additionally, proteins are reduced during equilibration to ensure their linearization and alkylated to prevent reformation of cysteine bridges. Originally, alkylation was applied to quench excess reducing agent which caused the phenomena of "point streaking" when the SDS-PAGE gels were subsequently silver-stained [48]. Nowadays, the usage of iodoacetamide to block SH-groups is part of the standard protocol for protein identification by mass spectrometry to prevent formation of di-peptides.

The equilibration encompasses a 15 min reduction and 15 min alkylation step to improve transfer of proteins into the second dimension [48]. It has been argued that the usage of HED while IEF does not require the application of a reduction/alkylation step while equilibration [49], but proper equilibration time still needs to be attended to. It should be noted that the mass of HED adduct to the cysteine groups of proteins needs to be specified in the peptide identification algorithms when conducting database searches of MS data for protein identification. After equilibration, the IPG strips are placed upon the SDS-PAGE gels. Two different approaches of SDS-PAGE are available, vertical and horizontal systems.

In case of vertical systems (e.g. Ettan DALT (GE Healthcare)), the gels are scanned while they are still in the cassette, so it is mandatory to use low-fluorescent glass cassettes. The headspace of the gel cassette needs to be washed to remove remnant overlay solution from gel casting and subsequently all liquid has to be removed from the gel surface. An easy way to achieve this is to tilt the gel cassette and use a tissue paper to remove remnant liquid without touching the gel surface. Application of the IPG strip onto the gel surface is less challenging when the strip is briefly dipped into "upper chamber" running buffer just beforehand as this (most of the times) prevents the strip from sticking to the glass plate. Make sure to touch only the plastic backing side when handling IPG strips. Place the strip in a slight upwards angle, this helps air-bubbles to escape when the agarose sealant is applied on top. Molecular weight markers can be easily pipetted onto a piece of filter paper and placed next to the IPG strip. Be aware that if there was remaining liquid on the gel surface from the washing steps, this will cause the molecular weight marker to leak out of the filter paper over the whole length of the gel surface. As a final step before the electrophoresis, the IPG strips are overlaid with agarose solution to keep them in place. The headspace of the glass cassettes should be filled completely. It is mandatory that all air-bubbles between the IPG strip and the gel surface are removed to prevent formation of "noses" and potential protein spot-splitting on the SDS-PAGE gel. The SDS-PAGE progress marker bromophenol blue can either be mixed into the equilibration solution or the agarose overlay solution (in case of mixing into the agarose solution, a slight blueish hue is sufficient). Electrophoresis should be started using a weak electric field, application of 5 mA per gel for an hour (using large format (20×25 cm^2) gels) produces an electric field sufficient to facilitate a smooth protein transfer from the IPG strip into the SDS-PAGE gel. Afterwards the current is increased to 8 mA per gel for an hour, followed by 15–20 mA per gel until the bromophenol blue front reaches to approx. 1 cm from the lower gel edge (this usually takes overnight). As for IEF, SDS-PAGE needs to be conducted in the dark to prevent dye bleaching.

In case of horizontal systems (e.g. HPE tower system (Serva)), the polyacrylamide gels are not in cassettes but are cast onto a (low-fluorescent) plastic backing. The gels are placed onto a cooling plate, a liquid facilitates homogenous heat transfer between the gel and a cooling device. It is therefore mandatory to remove all air-bubbles between the gel and the cooling plate as heat-pockets will influence the electrophoretic behavior in their proximity. The IPG strip is placed with the gel surface down onto the gel and air-bubbles are gently removed. Overlay with agarose is not necessary and as creating a gel/agarose overlay with inhomogeneous surface thickness and associated influence of the electric field, is discouraged. Transfer of the proteins from the IPG strips into the gel follows the same principle as described in the vertical system section; however, the final electrophoresis step can be conducted at

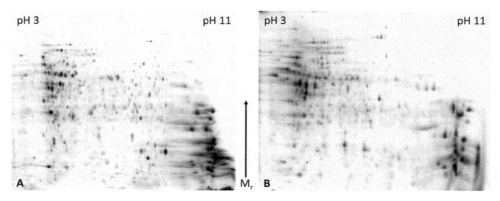

Fig. 2 Improved visualization of lower MW proteins in second dimension by using a **(a)** flatbed horizontal system compared to **(b)** a vertical electrophoresis system. From: Arentz, G.; Weiland, F.; Oehler, M. K.; Hoffmann, P., State of the art of 2D DIGE. *Proteomics Clin Appl* 2015, 9, (3-4), 277-88

higher electric field strengths as the gels are individually cooled. This enables a rapid finish of the electrophoresis within 5 h, and prevents lateral diffusion of low molecular weight proteins, enhancing their detection (*see* Fig. 2). As with horizontal systems, SDS-PAGE is conducted in the dark to prevent dye bleaching. The gels are scanned with the plastic backing on top and it is recommended to put some weight over the whole length of the upper and lower end of the gel backing to prevent curling up due to moisture evaporating while scanning.

2.6   Imaging and Data Analysis

Scanning DIGE gels digitalizes protein abundance information and enables data analysis using computers. Generally, the gels are digitized using a laser scanner or a high-resolution CCD camera. For a proper representation of the actual protein abundance on the gel, the whole dynamic range of the fluorescent dyes and scanner should be employed. Signal saturation of protein spots needs to be avoided as this influences subsequent intensity normalization using software tools. Gels should also be scanned with a high resolution (generally a 1 pixel per 100 μm (approx. 250 dpi) is used), while scanned files should only be saved in a loss-less format (e.g. TIFF). Our observation is that when the protein sample volumes have been equalized before the labeling reaction, the optimal photomultiplier tension (using a laser scanner) or exposure time (using a CCD camera) is the same for all gels throughout the series and needs to be set only once. After scanning, flatbed gels can be stored at −80 °C to preserve them for subsequent spot picking and protein identification. For horizontal systems, gels can either be removed from the glass cassettes or put into fixing solution. In case the cassettes have been prepared for spot picking by polymerizing one gel side onto one part of the glass cassette, the gels can be stored for up to a week at 4 °C while keeping them moist. This can easily be achieved by wrapping them into wet paper towels and plastic film.

Data analysis in DIGE encompasses (1) spot detection (2) quantification of either protein spot volume or peak height [50, 51], (3) correction of these values to account for systematic biases [52] and differences in extinction coefficients of the employed fluorescent dyes [53], (4) transformation of the data and (5) statistical analysis.

A pre-requisite for statistical analysis is the correct spot matching across the set of gels by stretching the gel images without altering spot volumes and intensities (warping). Commonly, two approaches are employed, warping before or after spot detection [54–56]. Warping before spot detection has the advantage of being able to integrate spot appearances throughout the experimental series without loss of data [50]. This approach also avoids the decreasing portion of accurately matched spots encountered with a warping after spot detection approach [57]. An easy, high-throughput capable gel warping algorithm for this is robust automated image normalization (RAIN), which can be downloaded for free [54, 55]. Protein quantification can either be done by spot boundary detection and volume calculation or by peak detection. Peak height strongly correlates to spot volume [51] and has been shown to be more reliable than volume calculations [50]. As a final step before commencing with statistical analysis, DIGE data is transformed to follow a normal distribution. This has the advantage that parametric statistical tests can be used, which are generally easier to interpret than nonparametric tests. Commonly applied multivariate analysis methods like PCA assume normally distributed data as well. Log_{10} transformation was originally suggested to transform DIGE data to follow a normal distribution [33], although this is not optimal as this approach introduces bias in the group of low intensity/volume spot group and does not eliminate a variance/mean dependence [53]. Several additional approaches have recently been described [53, 58]. While these novel approaches improve on simple log_{10} transformation, no clearly superior algorithm emerged from this set of alternative transformation methods and it was suggested to normalize using different approaches in parallel [58].

After transformation, a common method to compare two samples for differences in mean protein abundancies is a t-test. As univariate, parametric test the results are easy to interpret. A t-test calculates the probability that the two observed sample distributions are actually part of the same population distribution and is therefore a test for differences in the arithmetic mean of two sample populations. However, over recent years the usage of linear models, e.g. *limma* or Reproducibility-Optimized Test Statistic (ROTS), has emerged exhibiting a better statistical power [59].

However, these statistical procedures have only been tested in label-free MS data, so their capability in DIGE data would need to be evaluated first. As DIGE allows for the statistical testing of several thousand protein spots, the type I error needs to be controlled (as with all proteomics methods). Commonly applied is the q-value approach of Storey et al. [60, 61], an extension of the Benjamini and Hochberg method [62], which gives the minimum proportion of false positive discoveries (false discovery rate (FDR)) in the group of discoveries with a p-value smaller or equal to a p-value set by the researcher. In a multiple testing scenario it is therefore generally advisable to use a q-value cut-off rather than a p-value cut-off.

Multivariate statistics like PCA can give insight about clustering of samples, i.e. their similarity, and in combination with a loading plot, highlight the feature(s) most responsible for the samples or groups position on the PCA plot. However, multivariate methods are generally more difficult to interpret than univariate methods. More extensive information on statistical testing and modeling is beyond the scope of this chapter; the interested reader is referred to the "Handbook of Biological Statistics" by McDonald and "An Introduction to Statistical Learning" by James, Witten, Hastie and Tibshirani, both of which can be downloaded for free.

After deciding on the set of protein spots of interest as informed by statistical analysis, the protein spots are excised from the gel for protein identification by mass spectrometry. This can either be done manually after staining the gel with Coomassie Brilliant Blue or by employing a spot cutting robot (e.g. Ettan Spot Picker (GE Healthcare)). When employing a warping before spot detection strategy, the actual coordinates of the spot on the warped image will not correspond to the position on the gel, therefore the coordinates of spots of interest need to be assigned on the nonwarped gels. In case of low abundance proteins, pooling of corresponding spots from several gels is recommended.

3   Application of DIGE in Neuroproteomics

DIGE has been successfully applied to non-disease-related questions regarding brain development. A study by Laeremans et al. describes changes in protein abundance in mice 10 and 30 days after birth and adult mice [63]. Several hundred proteins were detected as differentially regulated, with the most striking changes between 10 days vs. 30 days postnatal and adult mice. Downregulation of six proteins implicated in Semaphorin signalling and downregulation of seven proteins involved in the ubiquitylation pathway was confirmed by in situ mRNA hybridization, from which the authors concluded that neurite outgrowth guidance and proteolysis are key functions in early brain development

[63]. A further study by Pinaud et al. examined the changes in protein abundance in songbirds when stimulated with conspecific songs [64]. Although no differential protein abundance in the caudomedial nidopallium could be detected after 5 min of continuous stimulation, 1 h and 3 h of continuous stimulus caused the regulation of proteins involved in calcium binding and neurotransmitter secretion. Differential regulation of proteins ZENK, SYN2, PKM2 and CALB2 were validated using immunocytochemistry [64].

The impact of sleep deprivation on cortical and thalamic synapses in mice was investigated by Simor et al. [65]. The experimental setup allowed for comparison of a sleep deprived vs. a control group as well as a group which was allowed to recover after sleep deprivation vs. a second control group. The majority of changes in protein abundance were detected in the parietal cortex (as opposed to thalamus), with Dpsyl2 and Hspa8 showing opposite regulation patterns between brain regions [65]. Further, in the parietal cortex Hspa8 exhibited an enduring increase in abundance even after recovery sleep [65]. The findings were interpreted to support the synaptic homeostasis hypothesis of sleep, due to the marked changes in the synaptic proteome detected in this study [65].

The effect of maternity on the hypothalamus and medial prefrontal cortex of rats was investigated in two studies by Udvari et al. and Völgyi et al. [66, 67]. Udvari et al. investigated changes to the synaptic proteome of hypothalami from 6 mother and 6 pup-deprived control rats. A decrease in abundance of C1qbp in non-pup-deprived mother rats was validated using Western blot. Subcellular localization of the C1qbp was confirmed almost exclusively to the mitochondria of cells, with the highest density in cells of the arcuate and ventromedial hypothalamic nuclei [67]. Bioinformatical analysis between C1qbp and maternal adaption hormones revealed a connection to 32 oestrogen protein interactors, 17 oxytocin interactors and 27 prolactin interactors [67]. Völgyi et al. applied a similar approach to study the effects of motherhood on rat medial prefrontal cortices [66]. Decrease of Alpha-crystallin B (ACB) chain in pup-deprived rats was confirmed by Western blotting and, as most differentially regulated proteins found in this study are associated with postpartum depression, suggested ACB as a potential marker for this condition [66].

The impact of binocular pattern deprivation (BD) onto the maturation of the primary visual cortex in cats was investigated by Laskowska-Macois et al. [68]. Differential proteome analysis of area 17 between normal control, 2-month and 4-month-old BD kittens was conducted. While a differential regulation of CRMP2 and CRMP4 between 2-month and 4-month-old kittens in normal control and BD groups was confirmed by Western blot, unfortunately no difference in abundance between normal control and BD kittens could be detected [68].

DIGE has further been commonly used in neurodegenerative disease research. A notable study by Osorio et al. investigated changes in mice hippocampi expressing the human APOE4 isoform, which is shown to correlate with poorer cognitive performance and increased synaptic deficits than mice expressing human APOE3 isoform of APOE [69], and is commonly implicated in Alzheimer's disease (AD). Analysis of 2D-DIGE gels revealed the differential regulation of two isoforms out of a four isoform cluster of mortalin, with a striking increase of isoform c by 14-fold abundance in APOE4 mice, while isoform d is upregulated 3.4 fold in APOE3 mice [69]. Phosphoprotein staining indicated isoform b to be phosphorylated in contrast to isoform a, c, and d [69]. This differential expression pattern of mortalin isoform c was reproduced in human samples. Isoform c was higher in abundance in hippocampi from patients with homozygous APOE3 as in patients with homozygous APOE4. Additionally, isoform c was higher abundant in AD patients with homozygous APOE3 as in comparison with brains from patients with homozygous APOE3 showing no signs of disease [69]. The findings were proposed to be part of a cellular defense mechanism against oxidative stress in AD [69].

Gillardon et al. reported the posttranslational modification of Grp75/mtHsp70 in Tg2576 mice overexpressing a mutant human amyloid precursor protein (APP) [19]. Grp75/mtHsp70 is a mitochondrial heat shock protein and is involved in mitochondrial protein translocation [70]. Therefore, mitochondria were characterized using a blue native/SDS-PAGE DIGE approach. Complex changes in the subunit composition of supercomplex $I + III_2$ from a mitochondrial preparation were detected and analyzed in combination with oxygen flux measurement, suggesting a complex I dysfunction in Tg2576 mice in concert with a decline in cerebral glucose utilization [19]. Further, changes in the synaptic mitochondrial proteome from APP/PS1 mice, carrying a human 695 amino acid isoform of APP and a mutant human PS1 (PS1-dE9) were investigated by Völgyi et al. [71]. Differential protein regulation between 3-, 6-, and 9-months-old mice APP/PS1 mice showing gradually increasing cognitive decline and control C57BL/6 mice was analyzed. Differentially regulated proteins Ethe1 and Htra2 as detected by DIGE were validated by Western blot to be upregulated in APP/PS1 mice as compared with normal controls of same age and also a differential expression of both proteins in the 6-months-old group (as compared to the 3- and 9-months groups). The authors suggested the potential usefulness of both proteins as progression markers for ß-amyloid accumulation [71].

Laramée et al. studied the long-term effects of early visual loss on the proteome of the V1, V2M, and V2L regions of the mouse brain. The experimental design included normal sighted controls, mice bilaterally enucleated 24 h postnatal, and congenitally anophthalmic mice [72]. Twenty-five protein spots corresponding to 29 proteins were found to be differentially regulated ($p < 0.05$) between sighted and blind mice [72]. CRMP2 and CRMP4 were validated using Western blots to be upregulated in V2M region of blind mice as compared to controls, CRMP4 was further validated to be upregulated in the V1 of anophthalmic mice as compared to enucleated and control group mice [72].

Involvement of differential protein regulation in depression was investigated by Gellén et al. [73]. Rat pups were chronically treated with clomipramine (CLO) and the proteome of the prefrontal cortex was compared against a control group [73]. Macrophage migration inhibitory factor was detected to be upregulated in CLO treated mice and validated by Western blot [73].

4 Conclusions

DIGE is a mature method to quantify differential protein abundance: The method is sensitive, accurate, and reproducible and allows straightforward detection of differences in protein abundance across a set of samples. Although the throughput of the method is lower than mass-spectrometry using sample labeling (e.g. tandem-mass tags), DIGE as a top-down method retains vital information on protein isoform status as well as posttranslational modifications (proteoform). This capability is the striking advantage of 2D-DIGE as it takes into account the actual state of the differentially regulated proteins. Therefore, 2D gel electrophoresis can and should be used as a "pre-fractionation" method for mass spectrometric analysis, providing the purification and enrichment of proteins of interest. These proteoforms can then be studied in depth by mass spectrometry to shed light onto the posttranslational status of the players involved in activated biochemical pathways. Theoretically this can also be achieved by mass spectrometry, but top-down quantitative approaches of complex samples are still far from being a routine methods [74] and the complex and expensive instrumentation renders it an untenable option for smaller labs. Despite the many advantages, DIGE is a demanding method in terms of manual dexterity, and while performing 2D gel electrophoresis one can encounter many pitfalls (for an in-depth trouble shooting guide *see* [47]). Sample preparation needs to be optimized individually for every sample type to avoid problems especially during IEF. Furthermore, it is recommended to standardize every step of DIGE to avoid masking of biological variance by inflated technical variance. As with all methods employing multiple testing, DIGE is only a hypothesis generating tool, and all proteins of interest need to be validated using a different method (e.g. Western blotting) in an independent set of samples.

Acknowledgments

The author wants to thank Maithili Shroff for proof-reading the manuscript.

References

1. Unlu M, Morgan ME, Minden JS (1997) Difference gel electrophoresis: a single gel method for detecting changes in protein extracts. Electrophoresis 18(11):2071–2077
2. Alban A, David SO, Bjorkesten L, Andersson C, Sloge E, Lewis S, Currie I (2003) A novel experimental design for comparative two-dimensional gel analysis: two-dimensional difference gel electrophoresis incorporating a pooled internal standard. Proteomics 3(1):36–44
3. Collier TS, Muddiman DC (2012) Analytical strategies for the global quantification of intact proteins. Amino Acids 43(3):1109–1117
4. Kim MS, Pinto SM, Getnet D, Nirujogi RS, Manda SS, Chaerkady R, Madugundu AK, Kelkar DS, Isserlin R, Jain S, Thomas JK, Muthusamy B, Leal-Rojas P, Kumar P, Sahasrabuddhe NA, Balakrishnan L, Advani J, George B, Renuse S, Selvan LD, Patil AH, Nanjappa V, Radhakrishnan A, Prasad S, Subbannayya T, Raju R, Kumar M, Sreenivasamurthy SK, Marimuthu A, Sathe GJ, Chavan S, Datta KK, Subbannayya Y, Sahu A, Yelamanchi SD, Jayaram S, Rajagopalan P, Sharma J, Murthy KR, Syed N, Goel R, Khan AA, Ahmad S, Dey G, Mudgal K, Chatterjee A, Huang TC, Zhong J, Wu X, Shaw PG, Freed D, Zahari MS, Mukherjee KK, Shankar S, Mahadevan A, Lam H, Mitchell CJ, Shankar SK, Satishchandra P, Schroeder JT, Sirdeshmukh R, Maitra A, Leach SD, Drake CG, Halushka MK, Prasad TS, Hruban RH, Kerr CL, Bader GD, Iacobuzio-Donahue CA, Gowda H, Pandey A (2014) A draft map of the human proteome. Nature 509(7502):575–581
5. Tazo Y, Hara A, Onda T, Saegusa M (2014) Bifunctional roles of survivin-DeltaEx3 and survivin-2B for susceptibility to apoptosis in endometrial carcinomas. J Cancer Res Clin Oncol 140:2027–2037
6. Arentz G, Weiland F, Oehler MK, Hoffmann P (2015) State of the art of 2D DIGE. Proteomics Clin Appl 9(3–4):277–288
7. Gorg A, Drews O, Luck C, Weiland F, Weiss W (2009) 2-DE with IPGs. Electrophoresis 30(Suppl 1):S122–S132
8. Tonge R, Shaw J, Middleton B, Rowlinson R, Rayner S, Young J, Pognan F, Hawkins E, Currie I, Davison M (2001) Validation and development of fluorescence two-dimensional differential gel electrophoresis proteomics technology. Proteomics 1(3):377–396
9. Shaw J, Rowlinson R, Nickson J, Stone T, Sweet A, Williams K, Tonge R (2003) Evaluation of saturation labelling two-dimensional difference gel electrophoresis fluorescent dyes. Proteomics 3(7):1181–1195
10. Sitek B, Potthoff S, Schulenborg T, Stegbauer J, Vinke T, Rump LC, Meyer HE, Vonend O, Stuhler K (2006) Novel approaches to analyse glomerular proteins from smallest scale murine and human samples using DIGE saturation labelling. Proteomics 6(15):4337–4345
11. Arnold GJ, Frohlich T (2012) 2D DIGE saturation labeling for minute sample amounts. Methods Mol Biol 854:89–112
12. Weiland F, Zammit CM, Reith F, Hoffmann P (2014) High resolution two-dimensional electrophoresis of native proteins. Electrophoresis 35(12–13):1893–1902
13. Schagger H, von Jagow G (1991) Blue native electrophoresis for isolation of membrane protein complexes in enzymatically active form. Anal Biochem 199(2):223–231
14. Altenhofer P, Schierhorn A, Fricke B (2006) Agarose isoelectric focusing can improve resolution of membrane proteins in the two-dimensional electrophoresis of bacterial proteins. Electrophoresis 27(20):4096–4111
15. Zammit CM, Weiland F, Brugger J, Wade B, Winderbaum LJ, Nies DH, Southam G, Hoffmann P, Reith F (2016) Proteomic responses to gold(iii)-toxicity in the bacterium Cupriavidus metallidurans CH34. Metallomics 8(11):1204–1216
16. Heinemeyer J, Scheibe B, Schmitz UK, Braun HP (2009) Blue native DIGE as a tool for comparative analyses of protein complexes. J Proteome 72(3):539–544
17. Peters K, Braun HP (2012) Comparative analyses of protein complexes by blue native DIGE. Methods Mol Biol 854:145–154
18. Reisinger V, Eichacker LA (2012) Native DIGE of fluorescent plant protein complexes. Methods Mol Biol 854:343–353
19. Gillardon F, Rist W, Kussmaul L, Vogel J, Berg M, Danzer K, Kraut N, Hengerer B (2007)

Proteomic and functional alterations in brain mitochondria from Tg2576 mice occur before amyloid plaque deposition. Proteomics 7(4):605–616

20. Mertins P, Yang F, Liu T, Mani DR, Petyuk VA, Gillette MA, Clauser KR, Qiao JW, Gritsenko MA, Moore RJ, Levine DA, Townsend R, Erdmann-Gilmore P, Snider JE, Davies SR, Ruggles KV, Fenyo D, Kitchens RT, Li S, Olvera N, Dao F, Rodriguez H, Chan DW, Liebler D, White F, Rodland KD, Mills GB, Smith RD, Paulovich AG, Ellis M, Carr SA (2014) Ischemia in tumors induces early and sustained phosphorylation changes in stress kinase pathways but does not affect global protein levels. Mol Cell Proteomics 13(7):1690–1704

21. Espina V, Edmiston KH, Heiby M, Pierobon M, Sciro M, Merritt B, Banks S, Deng J, VanMeter AJ, Geho DH, Pastore L, Sennesh J, Petricoin EF 3rd, Liotta LA (2008) A portrait of tissue phosphoprotein stability in the clinical tissue procurement process. Mol Cell Proteomics 7(10):1998–2018

22. Grassl J, Westbrook JA, Robinson A, Boren M, Dunn MJ, Clyne RK (2009) Preserving the yeast proteome from sample degradation. Proteomics 9(20):4616–4626

23. Ahmed MM, Gardiner KJ (2011) Preserving protein profiles in tissue samples: differing outcomes with and without heat stabilization. J Neurosci Methods 196(1):99–106

24. Kanshin E, Tyers M, Thibault P (2015) Sample collection method bias effects in quantitative Phosphoproteomics. J Proteome Res 14(7):2998–3004

25. Smejkal GB, Rivas-Morello C, Chang JH, Freeman E, Trachtenberg AJ, Lazarev A, Ivanov AR, Kuo WP (2011) Thermal stabilization of tissues and the preservation of protein phosphorylation states for two-dimensional gel electrophoresis. Electrophoresis 32(16):2206–2215

26. Acosta-Martin AE, Chwastyniak M, Beseme O, Drobecq H, Amouyel P, Pinet F (2009) Impact of incomplete DNase I treatment on human macrophage proteome analysis. Proteomics Clin Appl 3(10):1236–1246

27. Righetti PG, Chiari M, Gelfi C (1988) Immobilized pH gradients: effect of salts, added carrier ampholytes and voltage gradients on protein patterns. Electrophoresis 9(2):65–73

28. Stark GR, Stein WH, Moore S (1960) Reactions of the Cyanate present in aqueous urea with amino acids and proteins. J Biol Chem 235(11):3177–3181

29. Rai AJ, Gelfand CA, Haywood BC, Warunek DJ, Yi J, Schuchard MD, Mehigh RJ, Cockrill SL, Scott GB, Tammen H, Schulz-Knappe P, Speicher DW, Vitzthum F, Haab BB, Siest G, Chan DW (2005) HUPO plasma proteome project specimen collection and handling: towards the standardization of parameters for plasma proteome samples. Proteomics 5(13):3262–3277

30. Viswanathan S, Unlu M, Minden JS (2006) Two-dimensional difference gel electrophoresis. Nat Protoc 1(3):1351–1358

31. Mujumdar RB, Ernst LA, Mujumdar SR, Waggoner AS (1989) Cyanine dye labeling reagents containing isothiocyanate groups. Cytometry 10(1):11–19

32. Karp NA, Lilley KS (2005) Maximising sensitivity for detecting changes in protein expression: experimental design using minimal CyDyes. Proteomics 5(12):3105–3115

33. Karp NA, Kreil DP, Lilley KS (2004) Determining a significant change in protein expression with DeCyder during a pair-wise comparison using two-dimensional difference gel electrophoresis. Proteomics 4(5):1421–1432

34. Lilley KS, Friedman DB (2004) All about DIGE: quantification technology for differential-display 2D-gel proteomics. Expert Rev Proteomics 1(4):401–409

35. Karp NA, McCormick PS, Russell MR, Lilley KS (2007) Experimental and statistical considerations to avoid false conclusions in proteomics studies using differential in-gel electrophoresis. Mol Cell Proteomics 6(8):1354–1364

36. Buncel E, Symons EA (1970) The inherent instability of dimethylformamide-water systems containing hydroxide ion. J Chem Soc D 3:164–165

37. Wang W, Ackermann D, Mehlich AM, Konig S (2011) Impact of quenching failure of cy dyes in differential gel electrophoresis. PLoS One 6(3):e18098

38. Wang W, Ackermann D, Mehlich AM, Konig S (2010) False labelling due to quenching failure of N-hydroxy-succinimide-ester-coupled dyes. Proteomics 10(7):1525–1529

39. Riederer IM, Herrero RM, Leuba G, Riederer BM (2008) Serial protein labeling with infrared maleimide dyes to identify cysteine modifications. J Proteome 71(2):222–230

40. Qu Z, Meng F, Zhou H, Li J, Wang Q, Wei F, Cheng J, Greenlief CM, Lubahn DB, Sun GY, Liu S, Gu Z (2014) NitroDIGE analysis reveals inhibition of protein S-nitrosylation

by epigallocatechin gallates in lipopolysaccharide-stimulated microglial cells. J Neuroinflammation 11:17

41. Poschmann G, Grzendowski M, Stefanski A, Bruns E, Meyer HE, Stuhler K (2015) Redox proteomics reveal stress responsive proteins linking peroxiredoxin-1 status in glioma to chemosensitivity and oxidative stress. Biochim Biophys Acta 1854(6):624–631

42. McNamara LE, Kantawong FA, Dalby MJ, Riehle MO, Burchmore R (2011) Preventing and troubleshooting artefacts in saturation labelled fluorescence 2-D difference gel electrophoresis (saturation DiGE). Proteomics 11(24):4610–4621

43. Bjellqvist B, Ek K, Righetti PG, Gianazza E, Gorg A, Westermeier R, Postel W (1982) Isoelectric focusing in immobilized pH gradients: principle, methodology and some applications. J Biochem Biophys Methods 6(4):317–339

44. Righetti PG (1988) Isoelectric focusing as the crow flies. J Biochem Biophys Methods 16(2):99–110

45. Olsson I, Larsson K, Palmgren R, Bjellqvist B (2002) Organic disulfides as a means to generate streak-free two-dimensional maps with narrow range basic immobilized pH gradient strips as first dimension. Proteomics 2(11):1630–1632

46. Altland K, Becher P, Rossmann U, Bjellqvist B (1988) Isoelectric focusing of basic proteins: the problem of oxidation of cysteines. Electrophoresis 9(9):474–485

47. Görg A, Klaus A, Lück C, Weiland F, Weiss W (2007) Two-dimensional electrophoresis with immobilized pH gradients for proteome analysis: a laboratory manual. Technische Universität München, Freising-Weihenstephan, München

48. Görg A, Postel W, Weser J, Günther S, Strahler JR, Hanash SM, Somerlot L (1987) Elimination of point streaking on silver stained two-dimensional gels by addition of iodoacetamide to the equilibration buffer. Electrophoresis 8(2):122–124

49. Westermeier R, Görg A (2011) Two-dimensional electrophoresis in proteomics. In: Protein purification. John Wiley & Sons, Inc., Hoboken, pp 411–439

50. Morris JS, Clark BN, Wei W, Gutstein HB (2010) Evaluating the performance of new approaches to spot quantification and differential expression in 2-dimensional gel electrophoresis studies. J Proteome Res 9(1):595–604

51. Morris JS, Clark BN, Gutstein HB (2008) Pinnacle: a fast, automatic and accurate method for detecting and quantifying protein spots in 2-dimensional gel electrophoresis data. Bioinformatics 24(4):529–536

52. Sellers KF, Miecznikowski J, Viswanathan S, Minden JS, Eddy WF (2007) Lights, camera, action! Systematic variation in 2-D difference gel electrophoresis images. Electrophoresis 28(18):3324–3332

53. Jung K, Gannoun A, Sitek B, Meyer HE, Stühler K, Urfer W (2005) Analysis of dynamic protein expression data. RevStat: Statist J 3(2):99–111

54. Dowsey AW, Dunn MJ, Yang GZ (2008) Automated image alignment for 2D gel electrophoresis in a high-throughput proteomics pipeline. Bioinformatics 24(7):950–957

55. Dowsey AW, English J, Pennington K, Cotter D, Stuehler K, Marcus K, Meyer HE, Dunn MJ, Yang GZ (2006) Examination of 2-DE in the human proteome organisation brain proteome project pilot studies with the new RAIN gel matching technique. Proteomics 6(18):5030–5047

56. Veeser S, Dunn MJ, Yang GZ (2001) Multiresolution image registration for two-dimensional gel electrophoresis. Proteomics 1(7):856–870

57. Clark BN, Gutstein HB (2008) The myth of automated, high-throughput two-dimensional gel analysis. Proteomics 8(6):1197–1203

58. Keeping AJ, Collins RA (2011) Data variance and statistical significance in 2D-gel electrophoresis and DIGE experiments: comparison of the effects of normalization methods. J Proteome Res 10(3):1353–1360

59. Pursiheimo A, Vehmas AP, Afzal S, Suomi T, Chand T, Strauss L, Poutanen M, Rokka A, Corthals GL, Elo LL (2015) Optimization of statistical methods impact on quantitative proteomics data. J Proteome Res 14(10):4118–4126

60. Storey JD, Tibshirani R (2003) Statistical significance for genomewide studies. Proc Natl Acad Sci U S A 100(16):9440–9445

61. Storey JD (2002) A direct approach to false discovery rates. J R Statist SocB 64(3):479–498

62. Benjamini Y, Hochberg Y (1995) Controlling the false discovery rate: a practical and powerful approach to multiple testing. J R Statist SocB 57:289–300

63. Laeremans A, Van de Plas B, Clerens S, Van den Bergh G, Arckens L, Hu TT (2013) Protein expression dynamics during postnatal mouse brain development. J Exp Neurosci 7:61–74

64. Pinaud R, Osorio C, Alzate O, Jarvis ED (2008) Profiling of experience-regulated pro-

teins in the songbird auditory forebrain using quantitative proteomics. Eur J Neurosci 27(6):1409–1422

65. Simor A, Gyorffy BA, Gulyassy P, Volgyi K, Toth V, Todorov MI, Kis V, Borhegyi Z, Szabo Z, Janaky T, Drahos L, Juhasz G, Kekesi KA (2017) The short- and long-term proteomic effects of sleep deprivation on the cortical and thalamic synapses. Mol Cell Neurosci 79:64–80

66. Volgyi K, Udvari EB, Szabo ER, Gyorffy BA, Hunyadi-Gulyas E, Medzihradszky K, Juhasz G, Kekesi KA, Dobolyi A (2017) Maternal alterations in the proteome of the medial prefrontal cortex in rat. J Proteome 153:65–77

67. Udvari EB, Volgyi K, Gulyassy P, Dimen D, Kis V, Barna J, Szabo ER, Lubec G, Juhasz G, Kekesi KA, Dobolyi A (2017) Synaptic proteome changes in the hypothalamus of mother rats. J Proteome 159:54–66

68. Laskowska-Macios K, Nys J, Hu TT, Zapasnik M, Van der Perren A, Kossut M, Burnat K, Arckens L (2015) Binocular pattern deprivation interferes with the expression of proteins involved in primary visual cortex maturation in the cat. Mol Brain 8:48

69. Osorio C, Sullivan PM, He DN, Mace BE, Ervin JF, Strittmatter WJ, Alzate O (2007) Mortalin is regulated by APOE in hippocam-pus of AD patients and by human APOE in TR mice. Neurobiol Aging 28(12):1853–1862

70. Voos W, Rottgers K (2002) Molecular chaperones as essential mediators of mitochondrial biogenesis. Biochim Biophys Acta 1592(1):51–62

71. Volgyi K, Haden K, Kis V, Gulyassy P, Badics K, Gyorffy BA, Simor A, Szabo Z, Janaky T, Drahos L, Dobolyi A, Penke B, Juhasz G, Kekesi KA (2017) Mitochondrial proteome changes correlating with beta-amyloid accu-mulation. Mol Neurobiol 54(3):2060–2078

72. Laramee ME, Smolders K, Hu TT, Bronchti G, Boire D, Arckens L (2016) Congenital Anophthalmia and binocular neonatal Enucleation differently affect the proteome of primary and secondary visual cortices in mice. PLoS One 11(7):e0159320

73. Gellen B, Volgyi K, Gyorffy BA, Darula Z, Hunyadi-Gulyas E, Baracskay P, Czurko A, Hernadi I, Juhasz G, Dobolyi A, Kekesi KA (2017) Proteomic investigation of the prefron-tal cortex in the rat clomipramine model of depression. J Proteome 153:53–64

74. Catherman AD, Skinner OS, Kelleher NL (2014) Top down proteomics: facts and per-spectives. Biochem Biophys Res Commun 445(4):683–693

Chapter 14

Neuropeptidomics of the Mammalian Brain

Fang Xie, Krishna D. B. Anapindi, Elena V. Romanova, and Jonathan V. Sweedler

Abstract

A suite of bioactive peptides orchestrates a variety of cellular interactions in the mammalian brain. The bioanalytical strategy known as neuropeptidomics evolved from the quest to globally characterize these important cell–cell signaling peptides. The goal of a neuropeptidomics experiment is to characterize the peptides present in an intact brain, brain region, or even an individual neuron. A neuropeptidomics measurement needs to contend with the large dynamic range and low abundance of the neuropeptides that are present within a background of peptides resulting from the postmortem degradation of high-level ubiquitous proteins. The core components of a successful effort include effective tissue sampling and stabilization, sensitive and robust peptide characterization, and comprehensive data analysis and interpretation. Mass spectrometry (MS) has become the central analytical approach for high-throughput characterization of the brain peptidome because of its capability to detect, identify, and quantify known and unknown peptides with high confidence. Robust fractionation techniques, such as two-dimensional liquid chromatography (LC), are commonly used in conjunction with MS to enhance investigation of the peptidome. Identification and characterization of peptides is more complex when neuropeptide prohormone genes have not been annotated in an unbiased manner. This chapter outlines techniques and describes protocols for three different experimental designs that combine MS with LC, each aimed at high-throughput discovery of peptides in brain tissue. Further, we describe the currently available bioinformatics tools for automatic query of the experimental data against existing protein databases, manual retrieval of structural information from raw MS data, and label-free quantitation.

Key words Neuropeptidome, Hormone, Neuropeptide, Bioinformatics, Liquid chromatography, Mass spectrometry

1   Introduction

Signaling peptides (SPs), which include many different types of endogenous peptides such as neuropeptides and peptide hormones, are involved in the regulation of various biological functions and behaviors [1–4]. Identifying the cell–cell SPs in tissues and cells provides a basis for understanding these numerous physiological processes and biological phenomena; however, the chemical analysis of SPs is challenging because of the inherent structural

Ka Wan Li (ed.), *Neuroproteomics*, Neuromethods, vol. 146,
https://doi.org/10.1007/978-1-4939-9662-9_14, © Springer Science+Business Media, LLC, part of Springer Nature 2019

and chemical complexity of the central nervous system, especially in mammals. Neuropeptides are produced via a series of enzymatic processing steps from protein precursors known as prohormones [5]. A single prohormone can encode multiple bioactive peptides, either replicates of an individual peptide, or single copies of different peptides, or both. Mass spectrometry (MS)-based platforms are well suited for the discovery and unambiguous characterization of bioactive peptides, as nearly all posttranslational modifications (PTMs) result in characteristic mass shifts.

Peptidomics, a term introduced by several groups in 2001 [6–8], is a field of study that aims to simultaneously identify the peptide complement of cells, tissues, organs, or an entire organism. Neuropeptidomics refers to the detailed analysis of endogenous peptides from the brain or other neuronal tissues, and can be considered a subfield of peptidomics. A typical peptidomics analysis of SPs usually consists of five major steps (*see* Fig. 1): (1) sample stabilization by chemical or heat denaturation of peptidases, (2) extraction of SPs from cells, tissues, or releasates, (3) separation by liquid chromatography (LC) or capillary electrophoresis (CE), (4) detection by MS, and finally, (5) identification of SPs via bioinformatics-assisted MS data mining [1, 2, 9, 10]. This chapter outlines the most common and effective MS-based protocols for neuropeptidomics studies published in the literature [11–16].

Sample preparation is a crucial step in the characterization of neuropeptides. No matter the technical figures of merit of the instrumentation used, if the methods used to sacrifice the animal and isolate the tissue are not done well, the measurements will not succeed. It is essential to minimize the postmortem proteolytic degradation of proteins and peptides, especially for mammalian brain samples. These proteolytic events can be delayed or minimized by

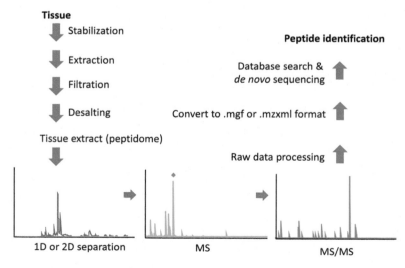

Fig. 1 Workflow of a typical peptidomics analysis

the use of protease inhibitor cocktails [17], instant freezing [18], heat inactivation of snap-frozen samples [18, 19], focused micro-wave irradiation [20], treating with hot water [3, 10], or by deacti-vating the enzymes by rapid conductive heating of the tissue at reduced pressures [11, 13]. Rapid conductive heat inactivation of tissues in particular has been shown to be more effective over other methods in preserving the endogenous complement of the pepti-dome [11]. Although both intact tissues and tissue homogenates have been used in neuropeptidomic studies, this chapter focuses on the latter. Multistage peptide extraction most often is based on the use of organic solvents in combination with strong acids. Methods that employ a series of buffers with different pH levels and polarities to improve the peptide's recovery from the sample have become popular as well [3, 11, 21]. Incorporating additional sample prepa-ration steps, such as desalting of the extracts, can also be helpful.

To reduce the complexity of the tissue homogenate or peptide extract, two-dimensional (2D) LC, at times at different scales in each dimension, is sometimes employed before the MS analysis. Two successive reversed-phase (RP) LC separations using different conditions [18], or using different modes (e.g., strong cation exchange (SCX) followed by RPLC [22]), have been used to sepa-rate neuropeptides in tissue extracts. The two parallel separations can be coupled either online or offline. As an example, two repre-sentative 2D LC setups are presented in this chapter: microbore RPLC offline coupled with nanoscale RPLC, and SCX online cou-pled with RPLC (*see* Fig. 2). The offline setup has the advantage of allowing large amounts of initial sample materials, as well as analysis of resulting fractions, using multiple MS platforms. Various LC-MS platforms have been employed in neuropeptidomic studies. Depending on the ion source, LC can be coupled to MS either online (via an electrospray ionization (ESI) source) or offline (via a

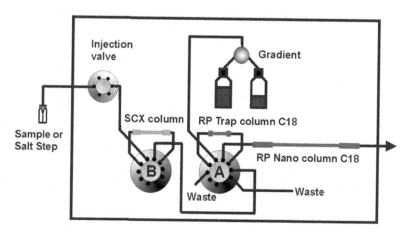

Fig. 2 Schematic diagram of the online SCX-RPLC setup. Adapted with permis-sion from ref. 2. Copyright © Wiley-VCH Verlag GmbH & Co. KGaA

matrix-assisted laser desorption/ionization (MALDI) source). Mass analyzers commonly used in neuropeptidomics include the ion trap (IT) [23–25], Fourier transform ion cyclotron resonance (FTICR) [24, 26], hybrid quadrupole orthogonal acceleration time-of-flight (Q-TOF) [22, 27, 28], TOF/TOF [10, 24, 29], and Orbitrap [27, 30]. Most of these systems are capable of alternating MS and tandem MS (MS/MS) scans automatically in a data-dependent manner. Here, we describe three LC-MS platforms that have been applied to neuropeptidomics: SCX-RPLC online coupled with ESI-Q-TOF MS, RPLC online coupled with ESI-Orbitrap MS, and RPLC offline coupled with MALDI-TOF/TOF MS.

Bioinformatic tools play a vital role in the processing of MS-derived data; in general, they help achieve two important objectives—the discovery/prediction of putative prohormone genes in newly sequenced genomes, and the identification of SPs to confirm/validate the predicted prohormones and peptides. The availability of genomic, transcriptomic, and proteomic information for a growing number of species allows in silico discovery of prohormones in newly sequenced species. Sequences of known prohormones from other species can be used to survey a nascent genome to find prohormone genes that were not previously known [31]. The construction of a neuropeptide prohormone database is the first step in peptide identification. The most likely set of putative SPs can be predicted from annotated prohormones using several binary logistic and expert system discovery tools, e.g., NeuroPred [32, 33]. A library of predicted peptides can aid in the follow-up MS analyses.

Bioinformatic algorithms identify peptides from their fragmentation spectra based on two basic principles: (1) spectrum match to an in silico peptide library (also known as a database search) and (2) de novo sequencing. In a database search, the algorithm identifies the peptide sequences by matching the generated MS/MS spectra to a list of peptides and proteins within a specified database. A variety of bioinformatic tools such as Mascot [34, 35], SEQUEST [36, 37], X!Tandem [38, 39], PEAKS Studio [9, 40], and Phenyx [41] have been developed to assist with this task. One major difference between a typical bottom-up proteomics experiment and a peptidomics analysis of endogenous peptides is the lack of an enzymatic digestion step during sample preparation. Therefore, the database search parameters should account for this difference and the "no enzyme" option chosen in the search parameter setup page found in most bioinformatics platforms. The different scoring algorithms employed by the above-named tools result in different probability scores from each platform, and careful attention must be given to interpreting the data and comparing the output from the various platforms utilized. In this chapter, we discuss the database search process using the PEAKS Studio software. However, other search engines should have a similar operational workflow.

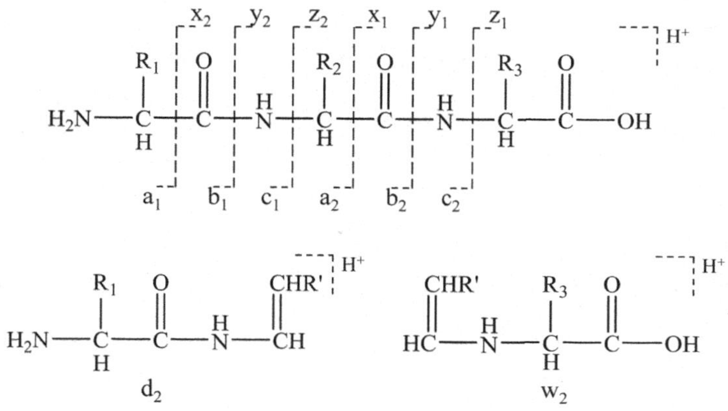

Fig. 3 Nomenclature of fragment ions generated in tandem mass spectrometry

For the MS/MS mass spectra that cannot be assigned via automatic protein/peptide database searches, de novo analyses of fragmentation patterns and characteristic mass shifts are performed. In a de novo approach for peptide sequencing, algorithms are used to construct a peptide sequence exclusively based on the fragmentation pattern of the obtained MS/MS spectra. The quality of the MS/MS data needed for de novo sequencing must be high. The fragmentation patterns and mass shifts characteristic of ion series from *N*- and/or *C*-termini are generated and serve to generate sequence tags for unassigned MS/MS mass spectra of putative peptides. The sequence tags obtained can then be used to survey the sequenced genome of the specific species being queried (which can be the first step in the discovery of novel prohormones), or to probe for prohormones in other species using homology searches. Some of the well-known de novo sequencing programs include SHERENGA [42], Lutefisk [43], NovoHMM [44], PepNovo [45] and pNovo [46]. This chapter describes the major steps of both de novo and database sequencing using collision-induced dissociation (CID) fragmentation spectra; MS/MS spectra obtained via other fragmentation mechanisms (*see* Fig. 3 and Table 1) can be interpreted similarly.

2   Materials

The materials needed and suppliers we use are listed here; comparable products from other suppliers should also be effective.

2.1   Sample Stabilization

1. The Stabilizor T1 (ST1) system (Denator, Uppsala, Sweden).
2. Maintainor Tissue cards (Denator).

2.2   Peptide Extraction

1. Methanol (MeOH; LC-MS grade; Thermo Fisher Scientific, Waltham, MA, USA).

Table 1
Fragment ion series of different types of MS instruments

	ESI-Q-TOF CID	ESI-IT/Orbitrap CID	ESI-IT/Orbitrap ETD	ESI-FTICR CID	ESI-FTICR ECD	MALDI-TOF/TOF CID
a ions						×
a-NH_3 if fragment includes RKNQ						×
a-H_2O if fragment includes STED						×
b ions	×	×		×		×
b-NH_3 if fragment includes RKNQ	×	×		×		×
b-H_2O if fragment includes STED	×	×		×		×
c ions			×		×	
y ions	×	×	×	×	×	×
y-NH_3 if fragment includes RKNQ	×	×		×		×
y-H_2O if fragment includes STED	×	×		×		×
z+H ions			×		×	
z+2H ions			×		×	
d or d' ions[a]						×
v ions						×
w or w' ions[a]						×

Adapted from the Matrix Science website (http://www.matrixscience.com/help/search_field_help.html)
[a]Isoleucine and threonine are doubly substituted at the beta carbon, so that side chain loss can give rise to two different ion structures. These pairs are designated d and d' or w and w'

2. Acetone, LC-MS grade (Thermo Fisher Scientific).

3. Water (H_2O; LC-MS grade, Thermo Fisher Scientific).

4. Hydrochloric acid (HCl; Thermo Fisher Scientific).

5. Microcon YM-10 filter unit (10 kDa molecular weight cut-off; MilliporeSigma, St. Louis, MO, USA).

2.3 Desalting

1. C18 ZipTip (MilliporeSigma), Pierce C18 Spin Column (MilliporeSigma), or Sep-Pak C18 cartridge (Waters, Milford, MA, USA).

2. CH_3CN, H_2O, formic acid (FA), trifluoroacetic acid (TFA) (LC-MS grade, Thermo Fisher Scientific).

2.4 Fractionation

1. Breeze HPLC system (Waters) with an RP C18 column (e.g., Vydac C18 MS column, 150 × 2.1 mm i.d., 5 μm particle size, 300 Å pores, Vydac, Hesperia, CA, USA).

2. UV detector.

3. Fraction collector.

4. MeOH (*see* Subheading 2.2).

5. HPLC solvents: Solvent A: 95% H_2O/5% CH_3CN/0.1% FA/0.01% TFA; Solvent B: 95% CH_3CN/5% H2O/0.1% FA/0.01% TFA (v/v/v/v, *see* Subheading 2.3).

6. Reconstitution solution: 95% H_2O/5% CH_3CN/0.01% TFA (v/v/v, *see* Subheading 2.3).

2.5 LC-MS Analysis

2.5.1 SCX-RPLC Online Coupled with ESI-Q-TOF MS

1. Ammonium acetate (MilliporeSigma).

2. H_2O (*see* Subheading 2.2 or 2.3).

3. RPLC solvents A and B (*see* Subheading 2.4).

4. Dionex UltiMate 3000 Proteomics MDLC system with dual ternary gradient pump (Thermo Fisher Scientific).

5. Dionex Ultimate 3000 Automated Fraction Collector (Thermo Fisher Scientific).

6. SCX column (e.g., PolySULFOETHYL aspartamide column, 15 cm × 300 μm i.d., Thermo Fisher Scientific).

7. C18 pre-column (e.g., RP Trap Cartridge, Thermo Fisher Scientific).

8. Analytical C18 column (e.g., PepMap C18, 150 mm × 75 μm i.d., 3 μm particle size, 100 Å pores, Thermo Fisher Scientific).

9. Impact HD UHR qQ-TOF mass spectrometer (Bruker, Billerica, MA, USA).

2.5.2 RPLC Online Coupled with ESI-Orbitrap MS

1. Dionex Ultimate 3000 RSLCnano system (Thermo Fisher Scientific) with an RP C18 column (e.g., PepMap C18, 150 mm × 300 μm i.d., 3 μm particle size, 100 Å pores, Thermo Fisher Scientific).

2. RPLC solvents should be different from those used in the first dimension to achieve better separation efficiency. If solvents A and B in the first dimension are as listed under Subheading 2.4, solvents A and B in the second dimension can be 98% H_2O/2% MeOH/0.01% heptafluorobutyric acid (HFBA, Thermo Fisher Scientific) and 98% MeOH/2% H_2O/0.01% HFBA, respectively.

3. ESI-Orbitrap mass spectrometer (e.g., Orbitrap Fusion Tribrid, Thermo Fisher Scientific).

2.5.3 RPLC Offline Coupled with MALDI-TOF/ TOF MS	1. Capillary LC system with an RP C18 column (*see* Subheading 2.5.2).

1. Capillary LC system with an RP C18 column (*see* Subheading 2.5.2).

2. RPLC solvents (*see* Subheading 2.5.2).

3. α-cyano-4-hydroxycinnamic acid (CHCA) or 2,5-dihydroxybenzoic acid (DHB) (MilliporeSigma).

4. Acetone, CH_3CN, H_2O, TFA (*see* Subheading 2.3).

5. Automated fraction collector/spot picker (e.g., ProteineerFC, Bruker).

6. MALDI-TOF/TOF mass spectrometer (e.g., ultrafleXtreme, Bruker).

2.6 Data Analysis

2.6.1 Peptide Identification with the Assistance of Bioinformatic Tools (Automatic De Novo and Database Search)

1. Software program for data format conversion (e.g., DataAnalysis by Bruker or Xcalibur by Thermo Fisher Scientific).

2. A bioinformatics platform for de novo MS/MS data interpretation; e.g., SHERENGA [42], Lutefisk [43], NovoHMM [44], PEAKS studio (Bioinformatics Solutions Inc., Waterloo, ON, Canada), PepNovo [45] and pNovo [46], or database search; e.g., Mascot (http://www.matrixscience.com/), X!Tandem (http://www.thegpm.org/TANDEM/), SEQUEST (Thermo Fisher Scientific), PEAKS Studio, and Phenyx (Geneva Bioinformatics (GeneBio) SA, Geneva, Switzerland).

3. Proteome database of the animal model of interest from a well-curated source. UniProt (https://www.uniprot.org/) and NCBI (https://www.ncbi.nlm.nih.gov) are good database download sources. Most database search programs accept the FASTA format.

2.6.2 Manual De Novo Sequencing and Subsequent BLAST Search

1. Basic Local Alignment Search Tool (BLAST; *see* **Note 1**) algorithms [31], either from the Baylor College of Medicine website (http://www.hgsc.bcm.tmc.edu/blast.hgsc), the National Center for Biotechnology Information (NCBI) website (http://blast.ncbi.nlm.nih.gov/Blast.cgi), or the standalone BLAST provided by NCBI (ftp://ftp.ncbi.nlm.nih.gov/blast/).

3 Methods

3.1 Sample Stabilization

1. The dissected tissue can either be directly transferred to the heat stabilization equipment or stored by snap freezing until later use. In either case, once the tissue is ready to be stabilized, carefully place it in the Maintainor Tissue card and close it with the lid facing up.

2. Process the tissue card in the ST1 machine according to the manufacturer's instructions. A few different setting combinations, as described by the manufacturer, accommodate most sample types.

3. Once the stabilization is complete, the tissue is removed from the Maintainor Tissue card and placed in the peptide extraction buffer.

3.2 Peptide Extraction

1. Homogenize the tissue sample in LC-MS grade water (pH 7) on a bed of ice (*see* **Note 2**) and let it remain on ice for 1 h.

2. Spin the homogenate for 20 min at 20,000 × g and 4 °C in a microcentrifuge tube to pellet any solid material; transfer the supernatant to a new microcentrifuge tube.

3. Add the second buffer, acidified acetone (40:6:1 acetone/water/HCl) or acidified methanol (10% glacial acetic acid in methanol), to the solid pellet, vortex for 1 min, and let it remain on ice for 1 h.

4. Spin the homogenate for 20 min at 20,000 × g at 4 °C in a microcentrifuge tube to pellet any solid material; transfer the supernatant to the microcentrifuge tube containing the extraction buffer from the first stage. Pre-concentrate the sample using a vacuum concentrator until the peptides are nearly dry.

5. Add the third extraction buffer, ice cold 0.25% (v/v) acetic acid/H_2O, to the remaining solid pellet. Vortex for 1 min and let it remain on ice for 1 h.

6. Spin the homogenate for 20 min at 20,000 × g at 4 °C in a microcentrifuge tube to pellet any solid material. Transfer the supernatant to the dried peptide vial from **step 4** and vortex well.

7. Spin the sample for 10 min at 20,000 × g at 4 °C to pellet any solid material. If any pellet is visible, transfer the supernatant to a new microcentrifuge tube.

8. Filter the supernatant through a Microcon YM-10 unit to remove large proteins.

3.3 Desalting (See Note 3)

1. Activate the sorbent with 50% H_2O/50% CH_3CN/0.1% FA and 0.01% TFA (v/v/v/v).

2. Equilibrate the sorbent with 95% H_2O/5% CH_3CN/0.1% FA and 0.01% TFA (v/v/v/v).

3. Load the sample.

4. Wash with 95% H_2O/5% CH_3CN/0.1% FA, and 0.01% TFA (v/v/v/v).

5. Elute the peptides with 30% H_2O/70% CH_3CN/0.1% FA, and 0.01% TFA (v/v/v/v).

6. Remove the CH_3CN in a vacuum concentrator and reconstitute the peptides in a solution containing 95% H_2O/5% CH_3CN/0.1% FA, and 0.01% TFA (v/v/v/v). The volume of the reconstituted sample should be compatible with the size of the injection loop and the scale of the LC instrument being used to fractionate the sample.

3.4 Fractionation

1. Wash the injection loop multiple times with MeOH followed by H_2O.

2. Equilibrate the system to the required starting conditions (5% solvent B, optimal flow rate for the column to be used).

3. Load the sample onto the injection loop and switch the selection valve in-line with the column.

4. Separate the peptides on a RP C18 column with a gradient using solvents A and B. The gradient can stay isocratic at 5% B for 5 min to allow peptides to move from the injection loop to the column and bind to the solid phase, then increase to 50% B over 60 min to elute peptides off the column, ramp up quickly to 80% and stay at 80% of B for another 5 min to elute any hydrophobic compounds, then ramp down to 5% of B in the next 10 min and maintain isocratic conditions to equilibrate the column for the next sample or blank.

5. Monitor the separation with UV detection set at 214 nm.

6. Collect the eluent with a fraction collector either manually or automatically according to the manufacturer's instructions.

7. Pre-concentrate fractions by removing the CH_3CN in a vacuum concentrator and reconstitute the peptides in a solution containing 95% H_2O/5% CH_3CN/0.01% TFA (v/v/v). The volume of the reconstituted sample should be compatible with the size of the injection loop and the scale of the LC instrument being used to analyze the fractions (for example, 5–10 μL or 1–5 μL for a capillary or nanoflow LC-MS platform, respectively).

3.5 LC-MS Analysis

3.5.1 SCX-RPLC Online Coupled with ESI-Q-TOF MS

1. Switch the RP analytical column offline.

2. Switch the SCX column online with the RP trap column.

3. Load the sample. Most peptides are trapped on the SCX column, and those that do not interact with the SCX column are trapped on the RP trap column.

4. Switch the SCX column offline.

5. Rinse the RP trap column for 5 min to remove the salts.

6. Switch the RP trap column online with the RP analytical column.

7. Elute the peptides from the RP trap column and separate them on the RP analytical column using a linear gradient from 5% to 95% of solvent B.

8. Elute the peptides from the SCX column by injecting 20 μL of 20 mM ammonium acetate solution. Peptides are trapped on the RP trap column.

9. Repeat **steps 1–7** nine more times, but incorporate a change in **step 3**; instead of loading the sample, 20 μL of successive

concentrations (20, 50, 100, 200, 400, 600, 800, 1000, and 2000 mM) of ammonium acetate should be loaded in **step 3** to elute peptides stepwise from the SCX column.

10. Interface the LC eluent online with the ESI source.

11. Perform MS scans in the range of 300–1500 m/z.

12. Employ the data-dependent acquisition method to trigger the MS/MS scan. The most intense ion(s) in the MS scan are selected as parent ions for fragmentation, with the dynamic exclusion set to two spectra within 1 min. Preference is given to ions with two or more charges. The singly-charged ions can be excluded, if desired. The collision energy can be ramped for the most efficient and reproducible MS/MS fragmentation. The MS/MS scans are typically performed in the range of 50–2000 m/z.

3.5.2 Nano Flow RPLC Online Coupled with ESI-Orbitrap MS

1. Load 5 µL of a reconstituted fraction sample onto the C18 pre-column trap at a 15 µL/min flow rate of the loading solvent.

2. After sample loading (typically 3–4 min), switch the trap column in-line with the RP C18 analytical column.

3. Separate the peptides on a RP C18 analytical column with a gradient using solvents A and B. The gradient should be determined based on the elution profile in the first dimension.

4. Interface the LC eluent online with the ion source on an ESI-Orbitrap mass spectrometer.

5. The following steps are mostly the same as described in **steps 11** and **12** in Subheading 3.5.1. Most Orbitrap mass spectrometers have fragmentation modes in addition to the collision induced dissociation (CID). Electron transfer dissociation (ETD) and high energy collisional dissociation (HCD) are the two most popular fragmentation modes in addition to CID. These fragmentation modes can either be used individually or in combination of two or three, depending on experimental requirements.

6. Additional parameters related to the ion injection value, resolution of analysis, and collision energy for fragmentation can be chosen, depending on the type of sample being analyzed.

7. The m/z range values for MS and MS/MS scans can be used as provided in Subheading 3.5.1.

3.5.3 RPLC Offline Coupled with MALDI-TOF/TOF MS

1. The LC separation is best achieved on a capillary flow rate LC system operated at 1–5 µL/min. The separation method can utilize the same gradient as described in Subheading 3.3, except that the total separation time can be increased by 30% and the eluent directed to a MALDI target plate and collected auto-

matically at a rate of 30–60 s/spot, depending on the flow rate and the volume of the individual sample that can accommodated by the particular MALDI sample plate being used.

2. Add matrix solution (10 mg/mL CHCA in 50% acetone/50% H_2O/0.01% TFA (v/v/v) or DHB in 50% CH_3CN/50% H_2O/0.01% TFA (v/v/v)) to each spot; typically 1–2 μL should be sufficient. If the fractions are first collected in vials, mix 1 μL of each fraction with 1 μL of the matrix solution directly on the target plate.

3. Allow the spots to air dry.

4. Load the target plate into the mass spectrometer.

5. Calibrate the instrument with a standard peptide mixture in the positive-ion reflectron mode within a range of 550–6000 m/z (or any m/z range of interest).

6. Acquire mass spectra of each sample spot by shooting the laser at random portions on the spot, and sum the spectra from all laser shots to obtain a comprehensive peptide profile of each spot. Adjust the laser power during acquisition to gain the optimal mass resolution and sensitivity. Analysis can be performed automatically using the autoXecute function in the flexControl software (Bruker).

7. Change the instrument to MS/MS mode, and acquire MS/MS spectra on the peaks of interest. Usually, higher laser power is required for MS/MS analysis, and multiple fragmentation spectra should be summed to obtain a good spectrum.

3.6 Data Analysis

3.6.1 Peptide Identification with the Assistance of Bioinformatic Tools

1. Construct the prohormone database in FASTA format and import it into the bioinformatics platform. If a custom prohormone database is unavailable, a whole proteome database can also be used (*see* **Note 4**).

2. Process and convert the MS/MS data to a universal format, such as the Mascot generic file format (.mgf) or the peak list format (.pkl).

3. Import the converted data file into a bioinformatics platform.

4. Set the parameters in the bioinformatics platform, such as mass tolerance (*see* **Note 5**) and PTMs (*see* **Note 6**), for automatic de novo sequencing and database search.

5. Execute the automatic de novo sequencing and the subsequent database search on the MS/MS data file.

6. Manually verify all of the obtained peptide identities for accurate ion series, reasonable cleavage sites, and PTM identification (*see* **Notes 7** and **8**).

7. Depending on the goals of the study, the identified peptide sequences can be used for further quantitation purposes (*see* **Note 9**).

3.6.2 Manual De Novo Sequencing and Subsequent BLAST Search

1. Examine the low-mass region for the b_2-ion. The b_2-/a_2-ion pair, separated by 28 Da, serves as a guide for the b_2-ion (*see* Fig. 3).

2. Examine the high-mass region for the first identifiable y-ion (ideally y_{n-1}). The list of possible amino acid combinations derived from the b_2-ion limits the possible residues to consider.

3. Extend the y-ion series toward the low-mass region, and/or extend the b-ion series towards the high-mass region, depending on which ion series is predominant in the spectra. As a y-ion is identified, the low-mass region is examined to identify the corresponding b-ion, and vice versa.

4. Examine the ions that seem to be from the neutral losses, and check if each loss is compatible with the deduced amino acid residue (*see* **Note 8**).

5. Search all of the obtained sequence tags against various genomic information databases using BLAST algorithms. We find that at least five consecutive amino acids in the query sequence are needed in order to obtain reasonable hits, especially when searching against nucleotide databases.

6. Reconstruct the complete precursor by examining the nucleotides surrounding the active sequence for the presence of various gene features such as translation start-sites, exons and introns, as well as termination signals.

4 Notes

1. BLAST is a set of similarity search programs. Both the query and the targeted database can be nucleotide, translated nucleotide, or protein. The choice of BLAST program is based on the purpose of the search. For example, "tblastn" is used to search the translated nucleotide database using a protein query. The "BLAST Program Selection Guide" provided on the NCBI website (http://www.ncbi.nlm.nih.gov/blast/producttable.shtml) can help you select the right service for a particular search.

2. We find that the amount of water is crucial for effective peptide recovery and recommend using at least a 10× v/v excess of solvent relative to the estimated volume of tissue.

3. There are a number of commercially available products for effective desalting at different scales, many of which rely on the use of C18 resin, a common adsorbent in RP chromatography. Depending on the sample size, desalting can be performed with a C18 ZipTip (SigmaMillipore), Pierce C18 SpinColumn (SigmaMillipore), or a Sep-Pak C18 cartridge (Waters; available in various loading capacities and volumes). The major

steps are essentially the same and include activation and equilibration of the desalting sorbent, loading the sample and binding of peptides to the sorbent, washing out the salt and other components that do not bind to the sorbent, and finally, eluting peptides off the sorbent. Manufacturer's instructions provide the most effective protocols and helpful guidelines for each recommended product. The elution solvents may vary, but usually contain organic modifiers, such as CH_3CN or MeOH, with acidic additives such as FA and/or TFA.

4. In a typical database search, a higher number of peptide spectrum matches cross a specified false discovery rate threshold with a targeted protein database rather than a more generic database that contains predominantly irrelevant proteins.

5. Mass tolerance depends on the instrument used for data acquisition. For example, the mass tolerance can be set at 2–5 ppm (precursor ions) and 10 ppm (fragment ions) for Orbitrap MS, but set at 10 ppm (precursor ions) and 40 ppm (fragment ions) for Q-TOF MS.

6. Common modifications include C-terminal amidation, formation of N-terminal pyroglutamic acid from N-terminal Glu and Gln, and disulfide bonding.

7. The C-terminal amide group is known to be derived from a Gly residue preceding the cleavage sites in the precursor. If the software assigned a C-terminally amidated peptide without a Gly residue, followed by the prohormone sequence, this assignment would not be correct. In addition, a minimum of three consecutive ion (b- and y-ion) matches is required to be a true-positive match.

8. Water loss (−18 Da) is expected for Ser, Thr, Asp, and Glu; ammonium loss (−17 Da) is expected for Asn, Gln, Lys, and Arg. A loss of 48 Da may be observed for Met, or −64 Da if it is oxidized. Table 1 provides a comprehensive list of expected fragment ions from each MS platform.

9. The peptides identified from the database search can further be used for label-free relative quantitation. Software tools such as Skyline [47] (http://www.skyline.ms), MaxQuant [48] (http://www.biochem.mpg.de/5111795/maxquant), and Mascot (http://www.matrixscience.com/) are some of the widely used software tools for label-free peptide quantitation. While Skyline and MaxQuant match the peptide identifications to the corresponding MS^1 extracted ion chromatograms (EIC) and calculate the area under EIC for quantitation, Mascot performs the quantitation directly from the MS/MS data by comparing the spectral counts of each peptide. These tools match the parent ion mass, isotopic pattern, and retention time of the library peptides with the sample to confirm identity.

Acknowledgments

This work was supported by Award Number P30 DA018310 from the National Institute on Drug Abuse (NIDA), and by Award No. NS031609 from the National Institute of Neurological Disorders and Stroke (NINDS). The content is solely the responsibility of the authors and does not necessarily represent the official views of the funding agencies.

References

1. Li L, Sweedler JV (2008) Peptides in the brain: mass spectrometry-based measurement approaches and challenges. Annu Rev Anal Chem 1(1):451–483. https://doi.org/10.1146/annurev.anchem.1.031207.113053

2. Boonen K, Landuyt B, Baggerman G, Husson SJ, Huybrechts J, Schoofs L (2008) Peptidomics: the integrated approach of MS, hyphenated techniques and bioinformatics for neuropeptide analysis. J Sep Sci 31(3):427–445. https://doi.org/10.1002/jssc.200700450

3. Lee JE, Zamdborg L, Southey BR, Atkins N, Mitchell JW, Li M, Gillette MU, Kelleher NL, Sweedler JV (2013) Quantitative peptidomics for discovery of circadian-related peptides from the rat suprachiasmatic nucleus. J Proteome Res 12:585–593. https://doi.org/10.1021/pr300605p

4. Zhang G, Vilim FS, Liu D-D, Romanova EV, Yu K, Yuan W-D, Xiao H, Hummon AB, Chen T-T, Alexeeva V, Yin S-Y, Chen S-A, Cropper EC, Sweedler JV, Weiss KR, Jing J (2017) Discovery of leucokinin-like neuropeptides that modulate a specific parameter of feeding motor programs in the molluscan model, Aplysia. J Biol Chem 292:18775. https://doi.org/10.1074/jbc.M117.795450

5. Rholam M, Fahy C (2009) Processing of peptide and hormone precursors at the dibasic cleavage sites. Cell Mol Life Sci 66:2075–2091. https://doi.org/10.1007/s00018-009-0007-5

6. Clynen E, Baggerman G, Veelaert D, Cerstiaens A, Van der Horst D, Harthoorn L, Derua R, Waelkens E, De Loof A, Schoofs L (2001) Peptidomics of the pars intercerebralis-corpus cardiacum complex of the migratory locust, Locusta migratoria. Eur J Biochem 268(7):1929–1939.https://doi.org/10.1046/j.1432-1327.2001.02067.x

7. Schrader M, Schulz-Knappe P (2001) Peptidomics technologies for human body fluids. Trends Biotechnol 19(10 Suppl):S55–S60. https://doi.org/10.1016/S0167-7799(01)01800-5

8. Verhaert P, Uttenweiler-Joseph S, de Vries M, Loboda A, Ens W, Standing KG (2001) Matrix-assisted laser desorption/ionization quadrupole time-of-flight mass spectrometry: an elegant tool for peptidomics. Proteomics 1(1):118–131. https://doi.org/10.1002/1615-9861(200101)1:1<118::AID-PROT118>3.0.CO;2-1

9. Zhang J, Xin L, Shan B, Chen W, Xie M, Yuen D, Zhang W, Zhang Z, Lajoie GA, Ma B (2012) PEAKS DB: de novo sequencing assisted database search for sensitive and accurate peptide identification. Mol Cell Proteomics 11:M111.010587. https://doi.org/10.1074/mcp.M111.010587

10. Tillmaand EG, Yang N, Kindt CAC, Romanova EV, Rubakhin SS, Sweedler JV (2015) Peptidomics and secretomics of the mammalian peripheral sensory-motor system. J Am Soc Mass Spectrom 26(12):2051–2061. https://doi.org/10.1007/s13361-015-1256-1

11. Yang N, Anapindi KDB, Romanova EV, Rubakhin SS, Sweedler J (2017) Improved identification and quantitation of mature endogenous peptides in the rodent hypothalamus using a rapid conductive sample heating system. Analyst 142:4476. https://doi.org/10.1039/C7AN01358B

12. Yin P, Hou X, Romanova EV, Sweedler JV (2011) Neuropeptidomics: mass spectrometry-based qualitative and quantitative analysis. Methods Mol Biol 789:223–236. https://doi.org/10.1007/978-1-61779-310-3_14

13. Sturm RM, Greer T, Woodards N, Gemperline E, Li L (2013) Mass spectrometric evaluation of neuropeptidomic profiles upon heat stabilization treatment of neuroendocrine tissues in crustaceans. J Proteome Res 12(2):743–752. https://doi.org/10.1021/pr300805f

14. Karlsson O, Kultima K, Wadensten H, Nilsson A, Roman E, Andrén PE, Brittebo EB (2013) Neurotoxin-induced neuropeptide perturbations in striatum of neonatal rats. J Proteome

Res   12(4):1678–1690.   https://doi.org/10.1021/pr3010265

15. Berezniuk I, Rodriguiz RM, Zee ML, Marcus DJ, Pintar J, Morgan DJ, Wetsel WC, Fricker LD (2017) ProSAAS-derived peptides are regulated by cocaine and are required for sensitization to the locomotor effects of cocaine. J Neurochem 143:268–281. https://doi.org/10.1111/jnc.14209

16. Romanova EV, Rubakhin SS, Ossyra JR, Zombeck JA, Nosek MR, Sweedler JV, Rhodes JS (2015) Differential peptidomics assessment of strain and age differences in mice in response to acute cocaine administration. J Neurochem 135(5):1038–1048.   https://doi.org/10.1111/jnc.13265

17. Garden RW, Shippy SA, Li L, Moroz TP, Sweedler JV (1998) Proteolytic processing of the Aplysia egg-laying hormone prohormone. Proc Natl Acad Sci U S A 95(7):3972–3977

18. Dowell JA, Heyden WV, Li L (2006) Rat neuropeptidomics by LC-MS/MS and MALDI-FTMS: enhanced dissection and extraction techniques coupled with 2D RP-RP HPLC. J Proteome Res 5(12):3368–3375. https://doi.org/10.1021/pr0603452

19. Colgrave ML, Xi L, Lehnert SA, Flatscher-Bader T, Wadensten H, Nilsson A, Andren PE, Wijffels G (2011) Neuropeptide profiling of the bovine hypothalamus: thermal stabilization is an effective tool in inhibiting post-mortem degradation. Proteomics 11(7):1264–1276. https://doi.org/10.1002/pmic.201000423

20. Che FY, Lim J, Pan H, Biswas R, Fricker LD (2005) Quantitative neuropeptidomics of microwave-irradiated mouse brain and pituitary. Mol Cell Proteomics 4(9):1391–1405. https://doi.org/10.1074/mcp.T500010-MCP200; [pii]: T500010-MCP200

21. Lee JE, Atkins N, Hatcher NG, Zamdborg L, Gillette MU, Sweedler JV, Kelleher NL (2010) Endogenous peptide discovery of the rat circadian clock: a focused study of the suprachiasmatic nucleus by ultrahigh performance tandem mass spectrometry. Mol Cell Proteomics 9(2):285–297. https://doi.org/10.1074/mcp.M900362-MCP200

22. Boonen K, Baggerman G, D'Hertog W, Husson SJ, Overbergh L, Mathieu C, Schoofs L (2007) Neuropeptides of the islets of Langerhans: a peptidomics study. Gen Comp Endocrinol 152(2-3):231–241. https://doi.org/10.1016/j.ygcen.2007.05.002; [pii]: S0016-6480(07)00174-8

23. Zhang X, Scalf M, Berggren TW, Westphall MS, Smith LM (2006) Identification of mammalian cell lines using MALDI-TOF and LC-ESI-MS/MS mass spectrometry. J Am Soc Mass Spectrom 17(4):490–499. https://doi.org/10.1016/j.jasms.2005.12.007; [pii]: S1044-0305(05)01046-9

24. Bora A, Annangudi SP, Millet LJ, Rubakhin SS, Forbes AJ, Kelleher NL, Gillette MU, Sweedler JV (2008) Neuropeptidomics of the supraoptic rat nucleus. J Proteome Res 7(11):4992–5003. https://doi.org/10.1021/pr800394e

25. Hook V, Bandeira N (2015) Neuropeptidomics mass spectrometry reveals signaling networks generated by distinct protease pathways in human systems. J Am Soc Mass Spectrom 26(12):1970–1980. https://doi.org/10.1007/s13361-015-1251-6

26. Su J, Sandor K, Sköld K, Hökfelt T, Svensson CI, Kultima K (2014) Identification and quantification of neuropeptides in naïve mouse spinal cord using mass spectrometry reveals [des-Ser1]-cerebellin as a novel modulator of nociception. J Neurochem 130(2):199–214. https://doi.org/10.1111/jnc.12730

27. Ye H, Wang J, Zhang Z, Jia C, Schmerberg C, Catherman AD, Thomas PM, Kelleher NL, Li L (2015) Defining the neuropeptidome of the spiny lobster Panulirus interruptus brain using a multidimensional mass spectrometry-based platform. J Proteome Res 14(11):4776–4791. https://doi.org/10.1021/acs.jproteome.5b00627

28. Fricker LD (2016) Proteolytic processing of neuropeptides. In: Grant JE, Li H (eds) Analysis of post-translational modifications and proteolysis in neuroscience. Springer New York, New York, NY, pp 209–220. https://doi.org/10.1007/7657_2015_87

29. Maki AE, Sweedler JV (2014) Characterizing neuropeptide release: from isolated cells to intact animals. In: Wilson GS, Michael AC (eds) Compendium of in vivo monitoring in real-time molecular neuroscience. World Scientific, Singapore, pp 335–349. https://doi.org/10.1142/9789814619776_0015

30. Saidi M, Beaudry F (2017) Targeted high-resolution quadrupole-Orbitrap mass spectrometry analyses reveal a significant reduction of tachykinin and opioid neuropeptides level in PC1 and PC2 mutant mouse spinal cords. Neuropeptides 65(Supplement C):37–44. https://doi.org/10.1016/j.npep.2017.04.007

31. Altschul SF, Lipman DJ (1990) Protein database searches for multiple alignments. Proc Natl Acad Sci U S A 87(14):5509–5513

32. Amare A, Hummon AB, Southey BR, Zimmerman TA, Rodriguez-Zas SL, Sweedler JV (2006) Bridging neuropeptidomics and genomics with bioinformatics: prediction of mammalian neuropeptide prohormone process-

ing. J Proteome Res 5(5):1162–1167. https://doi.org/10.1021/pr0504541

33. Southey BR, Amare A, Zimmerman TA, Rodriguez-Zas SL, Sweedler JV (2006) NeuroPred: a tool to predict cleavage sites in neuropeptide precursors and provide the masses of the resulting peptides. Nucleic Acids Res 34(Web Server issue):W267–W272. https://doi.org/10.1093/nar/gkl161; [pii]: 34/suppl_2/W267

34. Perkins DN, Pappin DJ, Creasy DM, Cottrell JS (1999) Probability-based protein identification by searching sequence databases using mass spectrometry data. Electrophoresis 20(18):3551–3567. https://doi.org/10.1002/(SICI)1522-2683(19991201)20:18<3551::AID-ELPS3551>3.0.CO;2-2

35. Barbé F, Le Feunteun S, Rémond D, Ménard O, Jardin J, Henry G, Laroche B, Dupont D (2014) Tracking the in vivo release of bioactive peptides in the gut during digestion: mass spectrometry peptidomic characterization of effluents collected in the gut of dairy matrix fed mini-pigs. Food Res Int 63(Part B):147–156. https://doi.org/10.1016/j.foodres.2014.02.015

36. Eng JK, McCormack AL, Yates JR (1994) An approach to correlate tandem mass spectral data of peptides with amino acid sequences in a protein database. J Am Soc Mass Spectrom 5(11):976–989. https://doi.org/10.1016/1044-0305(94)80016-2

37. Wu C, Monroe ME, Xu Z, Slysz GW, Payne SH, Rodland KD, Liu T, Smith RD (2015) An optimized informatics pipeline for mass spectrometry-based peptidomics. J Am Soc Mass Spectrom 26(12):2002–2008. https://doi.org/10.1007/s13361-015-1169-z

38. Craig R, Beavis RC (2004) TANDEM: matching proteins with tandem mass spectra. Bioinformatics 20(9):1466–1467. https://doi.org/10.1093/bioinformatics/bth092

39. Guerrero A, Dallas DC, Contreras S, Chee S, Parker EA, Sun X, Dimapasoc L, Barile D, German JB, Lebrilla CB (2014) Mechanistic peptidomics: factors that dictate specificity in the formation of endogenous peptides in human milk. Mol Cell Proteomics 13(12):3343–3351. https://doi.org/10.1074/mcp.M113.036194

40. Ma B, Zhang K, Hendrie C, Liang C, Li M, Doherty-Kirby A, Lajoie G (2003) PEAKS: powerful software for peptide de novo sequenc-

ing by tandem mass spectrometry. Rapid Commun Mass Spectrom 17(20):2337–2342. https://doi.org/10.1002/rcm.1196

41. Colinge J, Masselot A, Cusin I, Mahe E, Niknejad A, Argoud-Puy G, Reffas S, Bederr N, Gleizes A, Rey PA, Bougueleret L (2004) High-performance peptide identification by tandem mass spectrometry allows reliable automatic data processing in proteomics. Proteomics 4(7):1977–1984. https://doi.org/10.1002/pmic.200300708

42. Dančík V, Addona TA, Clauser KR, Vath JE, Pevzner PA (1999) De novo peptide sequencing via tandem mass spectrometry. J Comput Biol 6:327–342. https://doi.org/10.1089/106652799318300

43. Taylor JA, Johnson RS (2001) Implementation and uses of automated de novo peptide sequencing by tandem mass spectrometry. Anal Chem 73(11):2594–2604. https://doi.org/10.1021/ac001196o

44. Fischer B, Roth V, Roos F, Grossmann J, Baginsky S, Widmayer P, Gruissem W, Buhmann JM (2005) NovoHMM: a hidden Markov model for de novo peptide sequencing. Anal Chem 77:7265–7273. https://doi.org/10.1021/ac0508853

45. Frank A, Pevzner P (2005) PepNovo: de novo peptide sequencing via probabilistic network modeling. Anal Chem 77(4):964–973. https://doi.org/10.1021/ac048788h

46. Chi H, Chen H, He K, Wu L, Yang B, Sun R-X, Liu J, Zeng W-F, Song C-Q, He S-M, Dong M-Q (2013) pNovo+: de novo peptide sequencing using complementary HCD and ETD tandem mass spectra. J Proteome Res 12:615–625. https://doi.org/10.1021/pr3006843

47. MacLean B, Tomazela DM, Shulman N, Chambers M, Finney GL, Frewen B, Kern R, Tabb DL, Liebler DC, MacCoss MJ (2010) Skyline: an open source document editor for creating and analyzing targeted proteomics experiments. Bioinformatics 26:966–968. https://doi.org/10.1093/bioinformatics/btq054

48. Tyanova S, Temu T, Carlson A, Sinitcyn P, Mann M, Cox J (2015) Visualization of LC-MS/MS proteomics data in MaxQuant. Proteomics 15(8):1453–1456. https://doi.org/10.1002/pmic.201400449

Chapter 15

MALDI Imaging Mass Spectrometry: Neurochemical Imaging of Proteins and Peptides

Jörg Hanrieder, Henrik Zetterberg, and Kaj Blennow

Abstract

The central nervous system (CNS) constitutes the most intricate tissue in the human body. Neurological diseases, in particular, have a complex pathophysiology and are heterogeneous in their pathological and clinical presentation and therefore poorly understood on a molecular level. Increased insight in molecular CNS disease pathophysiology relates directly to the advancement of novel bioanalytical technologies that allow highly resolved, sensitive, specific, and comprehensive molecular analysis and molecular imaging in complex biological tissues, and in the CNS in particular. Imaging mass spectrometry (IMS) is an emerging technique for molecular imaging, characterized by its high molecular specificity and is therefore a powerful approach for investigating molecular localization patterns in CNS-derived tissue and cells. Over the last 20 years, IMS has been demonstrated to be a promising technology for chemical imaging in biochemical studies, but its application in clinical research is still in its infancy. The goal of this chapter is to provide the reader with a detailed step-by-step guide through the IMS workflow for the successful replication of published experimental data. Moreover, the aim is to give a concise overview of the major developments and applications of matrix-assisted laser desorption ionization (MALDI) based imaging mass spectrometry for neurochemical profiling with particular focus on protein and peptide imaging in neurodegenerative disease pathology.

Key words Imaging mass spectrometry (IMS), Matrix-assisted laser desorption ionization (MALDI), Central nervous system (CNS), Neurodegeneration, Alzheimer's disease (AD), Parkinson's disease (PD), Amyotrophic lateral sclerosis (ALS)

1 Introduction: Imaging Mass Spectrometry

With the average age of the world's population rising, the prevalence of age related diseases, including neurodegenerative diseases such as Alzheimer's disease (AD) and Parkinson's disease (PD) is on the march. Much details on the pathophysiological mechanisms underlying neurodegenerative disease pathology remain elusive, which in turn significantly hampers the development of curative treatment strategies. Some of the recent advances in the understanding of CNS-related molecular mechanisms in neurodegeneration have been made because of the development of novel,

Ka Wan Li (ed.), *Neuroproteomics*, Neuromethods, vol. 146,
https://doi.org/10.1007/978-1-4939-9662-9_15, © Springer Science+Business Media, LLC, part of Springer Nature 2019

advanced high-resolution imaging techniques. In modern neuro-biological research, chemical imaging techniques are essential bio-analytical tools to gain in depth understanding of molecular mechanism and interactions at the subcellular scale. Here different molecular imaging modalities involving, e.g., immunohistochem-istry, in situ hybridization, chemical probes and fluorescent micros-copy as well as spectroscopic methods are employed to retrieve topographical and temporal information of molecular abundance distributions. A major challenge in imaging lies on obtaining suit-able spatial resolution while maintaining high molecular specificity and sensitivity.

A rather novel molecular imaging technology, imaging mass spectrometry (IMS), has over the last years gained significant impact in biomedical research and neuroscience research in particular.

Due to the revolutionizing development of soft ionization techniques such as MALDI [1] and electrospray ionization (ESI) [2], mass spectrometry (MS) became the method of choice for protein and peptide characterization. MS allows fast, sensitive, and specific detection and characterization of intact large biomolecules including peptides and proteins. MS-based approaches of tissue extracts facilitate sensitive protein and peptide profiling; however, no accurate spatial information is maintained. Due to the complex-ity of the CNS, the spatial information of neurochemical distribu-tion patterns is of major interest in order to delineate ongoing neuronal mechanisms.

IMS is an emerging, novel approach for spatial mapping of chemical species in biological tissue and single cells [3, 4]. In con-trast to common, antibody- or nucleotide-based detection approaches (immunohistochemistry, in situ PCR), IMS does not require any a priori knowledge of the potential target species and is furthermore not dependent on antibody or primer availability and specificity.

IMS features high molecular specificity and allows comprehen-sive, multiplexed detection, identification, and localization of hun-dreds of proteins, peptides and lipids in biological tissue samples. Over the last decade, particularly MALDI IMS has slowly evolved as a relevant, alternative approach in biomedical research for study-ing proteins, peptides, lipids, drugs, and metabolites in disease pathology as well as in fundamental biological processes with par-ticular focus on neurological disorders [5, 6]. IMS enables to match histological features within a biological tissue sample to comprehensive molecular localization patterns [3, 7].

In general, IMS can be performed with different probes to desorb and ionize molecular species directly from a biological sam-ple (Fig. 1). The most prominent approaches along with MALDI IMS include time of flight secondary ion mass spectrometry (ToF-SIMS) [8], in which an ion beam is employed to sputter molecular

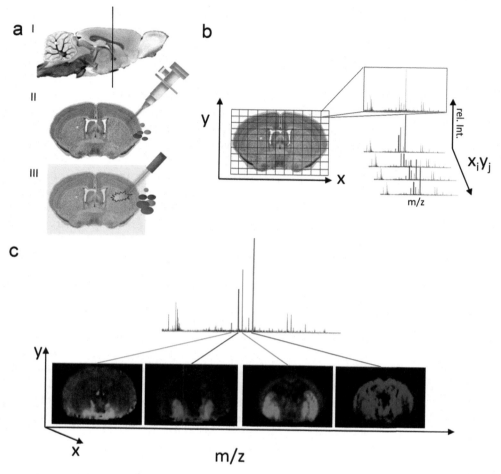

Fig. 1 Principle of imaging mass spectrometry. (**a**) I, Tissue sections are collected and mounted on a target for imaging MS. II, For SIMS IMS, tissue sections are probed with an ion beam, generating low molecular weight secondary ions (*m/z* > 1000 Da). III, In contrast, MALDI IMS requires precoating with matrix (indicated in yellow) before systematic scanning with a laser probe. MALDI-based ionization generates larger intact molecular species, including peptides and proteins. (**b**) One mass spectrum is acquired for every x_i, y_j coordinate of the tissue section based on a predefined array corresponding to pixel size and spatial resolution. (**c**) Single ion images are generated by mapping the intensity of an individual ion signal (*m/z*;rel.Int) over the whole tissue slide

species into the gas phase as well as desorption electrospray ionization (DESI) [9], where an electrospray for analyte desorption and ionization is focused onto the tissue. This method allows image data acquisition at atmospheric pressure though at larger spatial resolution.

For every IMS experiment, the molecular intensity distribution in biological samples is obtained by sequential acquisition of mass spectra from a predefined pixel array over a biological tissue section or a single cell. Single ion images are generated by plotting the intensity distribution of a distinct molecular species over the analyzed MS array (Fig. 1). The point-to-point distance defines the limit for lateral resolution of the IMS analysis.

Different IMS technologies have various strengths and limitations particularly with respect to spatial resolution and molecular information.

MALDI IMS represents a fair compromise between molecular mass range, spatial resolution, sensitivity, and molecular specificity. The technique involves laser irradiation-based desorption and ionization of molecular species with the help of a crystalline UV absorbing matrix [1]. The technique is characterized by its superior mass range, sensitivity, and mass accuracy at good spatial resolution as well its robustness, acquisition speed, and insensitivity to sample impurities. In 1997, a MALDI-based imaging approach was introduced that facilitated spatial profiling of large molecular species in mammalian tissue samples [10]. The MALDI IMS approach is based on application of a matrix layer onto a thawed mounted tissue section. The technique is particularly well suited for medium to large biomolecules including glycolipids, neuropeptides, and proteins.

The present methods chapter is therefore focusing on MALDI IMS and aims to provide a concise overview of its application for protein and peptide imaging in complex biological samples. Here, a step-by-step guide through the experimental setup and workflow is given to enable the reader to successfully replicate MALDI IMS experiments on site. Different advances as well as challenges with respect to sample preparation, data acquisition, and validation are discussed and finally an overview of IMS-based proteomic and peptidomic applications in biomedical neuroscience are provided.

2   Materials

2.1   Tissue Collection

1. Cryostat Microtome (Leica Biosystems, Nussloch, Germany).

2. Blades for cryostat (Leica Biosystems).

3. Indium tin oxide (ITO)-coated MALDI slides (Bruker Daltonics, Bremen, Germany).

4. Optimal cutting temperature medium (OCT) (Sakura Finetek Europe, The Netherlands).

5. Vacuum desiccator (VWR, Stockholm, Sweden).

2.2   Sample Preparation and Matrix Application

1. Acetonitrile (ACN), methanol (MeOH), ethanol (EtOH) of pro-analysis, trifluoroacetic acid (TFA), formic acid (FA) (Sigma Aldrich, St.Louis, MO, USA).

2. Double distilled water purified with a Milli-Q purification system (Merck-Millipore, Darmstadt, Germany).

3. 2,5-dihydroxyacetophenone (2,5-DHA, Sigma Aldrich)

4. TM Sprayer (HTX, Nashville, TN).

2.3   MALDI Imaging MS and Data Analysis

1. Ultraflextreme MALDI ToF/ToF intrument (Bruker Daltonics).

2. FlexImaging software (v 3.0, Bruker Daltonics).

3. Peptide and Protein Calibration Standard (Bruker Daltonics).

4. Origin v.8.5 (OriginLab Corp, Northampton, MA).

5. R (https://www.r-project.org).

6. GraphPad Prism (GraphPad Software, La Jolla, CA).

7. SCiLS Lab software v 2015 (Bruker Daltonics).

2.4   Immunohisto-chemistry

1. Chemicals: ethanol, acetone, PBS, Tween 20, triton-X (Sigma Aldrich).

2. Normal goat serum, BSA (Sigma Aldrich).

3. Primary antibody; mouse anti-human Aβ1–16 (6E10, BioLegend, San Diego, CA).

4. Secondary Antibody: rabbit or goat anti-mouse Alexa 480 (Invitrogen, Thermo Fisher Scientific, Carlsbad, CA, USA).

5. Prolong Gold antifade mountant (Thermo Fisher Scientific).

6. Wide-field microscope (Axio Observer Z1, Zeiss, Jena, Germany).

7. ImageJ software (http://rsb.info.nih.gov/ij/).

2.5   Tissue Proteomics and Peptidomics

1. Chemicals: Acetonitrile, acetone, methanol, of pro-analysis, acetic acid, trifluoroacetic acid, formic acid, SDS, dithiothreitol (DTT), iadoacetamide (Sigma Aldrich).

2. Double distilled water purified with a Milli-Q purification system (Merck-Millipore).

3. Trypsin (Promega, Madison, WI, USA).

4. Probe sonicator (Thermo Fisher Scientific, Waltham, MA, USA).

5. MWCO filter 30 kDa (Merck Millipore).

6. Normal flow, gradient HPLC System equipped with a well plate fraction collector for protein fractionation.

7. Nanoflow HPLC System equipped with loading pump, sample loop and trap column (Ultimate 3000, Thermo Scientific).

8. Reversed HPLC Columns: C8 (4.2 mm × 100 mm, Phenomenex) and C18 (0.075 mm × 150 mm, Acclaim PepMap, Thermo Scientific).

9. MS instrument (Orbitrap QExactive, Thermo Scientific) equipped with nano electrospray ESI source.

10. Excalibur software (Thermo Scientific).

11. Mascot software (MatrixScience, London, UK).

3 Methods

3.1 Sample Preparation for MALDI Imaging of Proteins and Peptides

For IMS of proteins and peptides, various target preparation parameters are of critical relevance as they have significant impact on final data quality. This is particularly with respect to signal intensity, reproducibility, as well as lateral (i.e. spatial) resolution. The sample preparation workflow in MALDI IMS comprises the tissue retrieval and tissue storage, tissue sectioning, tissue wash, and fixation followed by matrix application. In particular, tissue pre-treatment and matrix application are cornerstones for each MALDI IMS experiment, as they have to be adjusted for both the type of tissue and most importantly the molecular target of interest.

3.1.1 Tissue Collection

As for all tissue imaging experiments, tissue retrieval from animal or human sources is critical to data quality. This is of particular relevance, since commonly used perfusion and fixation strategies cannot be applied in mass spectrometry due to interference of the polymeric fixation agents such as paraffin or PFA. Elegant solutions to overcome these obstacles have been presented, including paraffin removal, antigen retrieval, and in situ trypsination [11, 12]. This approach, however, does not permit the analysis of intact, endogenous peptides, such as neuropeptides, due to the extensive washing for polymer removal and enzymatic degradation. Fresh frozen tissue samples are, therefore, the most preferred samples for IMS. In terms of sample dissection such as for intact brain, quick performance is essential, since postmortem delays by as little as 3 min result in severe degradation of neuropeptides [13]. One solution for overcoming this shortcoming, is heat stabilization, where quick sample heating to 95 °C leads to efficient protease inactivation [14]. This approach, however, impacts the sample morphology, which in turn hampers its suitability for IMS. This is primarily due to difficulties in maintaining tissue integrity during sectioning although an elegant workaround using collection on conductive carbon tape has been presented recently [15]. For fresh frozen tissue, cryosections are collected and thaw mounted onto conductive glass slides. Here, the collected sections have to be dried immediately before storage at −80 °C to prevent damage by water condensation during freezing [16].

3.1.2 Tissue Washing

A further critical step in sample handling is the choice of appropriate washing protocols. Lipid species typically do not require any advanced washing steps, whereas, drugs, neuropeptides and proteins require optimized washing protocols for signal enhancement. These involve pH sensitive clean up as well as sequential washes with organic solvents prior to matrix application for precipitation of peptides and proteins while washing off remaining lipids that potentially interfere with neuropeptide signals [16, 17]. Several washing protocols have been evaluated for enhancing protein signal in MALDI imaging. Stepwise washing with gradient alcohol has been found to give the most significant improvement in signal quality [18, 19].

3.1.3 Matrix Application

In terms of matrix application, there are currently three approaches available, each with its own strengths and limitations with respect to simplicity, cost, crystal size i.e. spatial resolution, and signal reproducibility. In general, there is a tradeoff between signal quality and lateral resolution based on the correlation of vertical and lateral diffusion with increased droplet size resulting in better extraction but bigger crystals and analyte delocation. The most straightforward approach involves manual application of the matrix solution employing an airbrush sprayer. While this is a quite cost-effective solution it is severely hampered by its lack in reproducibility as well its susceptibility to lateral sample diffusion as well as limited extraction efficiency. A more controlled way to deposit matrix is to use nebulizer-based sample application available in a commercial *ImagePrep* device (Bruker Daltonics, Bremen, Germany). Here, pneumatic aerosol formation and subsequent matrix application is monitored by a light scattering sensor below the sample glass to estimate matrix thickness. Although this is not as economical as the bare nebulizer, this technology has significantly advanced the aerosol-based matrix application in terms of reproducibility and signal quality. Further controlled means of nebulizer-based matrix deposition are found in the *TM sprayer* (HTX Technologies) or *SunCollect sprayer* (Sunchrome, Napa, CA, USA), where an nebulizer head is repeatedly moved over the tissue section thereby applying layer by layer of matrix. Most recently, a straightforward, cost efficient matrix deposition sprayer system was presented where all construction details are provided as open access (iMatrixSpray) [20].

Still, one major disadvantage of this solution remains in its rather low extraction efficiency, as a consequence of reduced lateral diffusion. The most controlled approach for sample application involves deposition by a chemical inkjet (*ChiP*, Shimadzu, Kyoto, Japan) or a pneumatic vertical spotter (*Portrait*, Labcite, Sunnyvale, CA). This approach uses accurate deposition of distinct droplets in a geometrical patter thereby avoiding lateral diffusion of the analyte [21]. The spotting approach is particularly beneficial for low abundant species, including neuropeptides [22]. Although this approach is limited in terms of spatial resolution, the best data quality in terms of signal strength and reproducibility is achieved.

Over the last years several alternative approaches for matrix application have been reported, to overcome the limitations in various matrix application protocols and to control crystal size, including dry coating, pre coating and sublimation [23, 24]. While, all of these techniques provide major advantages in terms of crystal size but are restricted due to their low extraction efficiency. Moreover, only sublimation has been reported for protein imaging, despite limitations with respect to reproducibility.

Several different matrices has been proposed for MALDI IMS of various molecular species. Different matrices are well suited for different compounds and the choice depends largely on the targeted substance. Many advances in matrix developments have been aimed at overcoming various limitations of the standard compounds, including crystal size, reproducibility, interfering matrix cluster, mass range, signal quality. For protein analysis, the most commonly used matrices are 2,5-dihydroxy-benzoic acid (DHB), sinnapinic acid (SA), 2,5-dihydroxy-acetophenone (2,5-DHA) and 4-hydroxyl-alpha-cyano-cinnaminic acid (HCCA).

3.2 Data Acquisition and Data Analysis

3.2.1 MALDI Imaging Analysis

The MALDI imaging experiment is detailed in a corresponding running sequence. Here, all the parameters of the acquisition and initial data processing are specified along with the acquisition region, acquisition pattern and spot-to-spot distance, i.e., spatial resolution. The initial steps of setting up the imaging acquisition experiment involve image-registration and teaching to fiducial markers , which is performed in the corresponding imaging software bundle provided by the vendor (e.g., FlexImaging, Bruker Daltonics).

For the actual MALDI acquisition method the optimum number of good-quality spectra that can be obtained from each position needs to be specified by adjusting the number of laser shots and the laser energy. Too high laser fluence can influence the experiment dramatically leading to signal depletion due to oversampling effects. Therefore, the spatial resolution together with the laser focus and laser energy settings have to be considered accordingly as to small spot-to-spot distances at too high laser fluence leads to the described oversampling phenomena and poor spatial resolution. This is particularly challenging for protein imaging experiments as typically higher laser fluences are needed for optimal desorption and ionization, thereby challenging the spatial resolution. For analysis ofa 2,5-DHA prepared tissue section using the above specified Ultraflextreme instrument, a number of 50 laser shots can typically be obtained from each spot at minimum laser focus for 20–30 μm pixel distance. Conversely, using 2,5-DHB or 1,5-DAN matrix for lipid analysis, only a number of 5–10 shots at lower energies are needed allowing straight forward acquisition of imaging data at 10 μm. By using a more recently introduced Rapiflex MALDI ToF instrument (Bruker Daltonics), robust and senitive protein imaging at 10µm is possible. A further consideration, when specifying the pixel distance and spatial resolution, respectively is the acquisition time. Assuming a MALDI instrument with a 1 kHz laser like the Ultraflextrem is used, acquisition of big areas at very high spatial resolution will take several hours. During this time the matrix might sublimate away rendering the data acquired later during the experiment worthless. Again, recent developments in laser interfaces such as for the Shimadzu 7090 (2kHz), SimulToF Two/Three (10kHz) and Bruker Rapiflex (10kHz) help to overcome this throughput issue.

3.2.2 Data Processing

Imaging data acquisition is generally followed by subsequent data processing including baseline subtraction, peak picking and calibration. On-the-fly MS processing is performed using baseline removal algorithm that will not result in negative data (e.g., ConvexHull 0.8 flatness; FlexAnalysis, Bruker Daltonics). Peak picking is performed using sensitive algorithms (Centroid, SNAP) considering peaks with a minimum signal to noise (S/N) >3.

Here, it is mandatory to always include at least three calibration spots placed in three corners diametrically over the slide and save the spectra in order to allow for revisiting the calibration and if necessary perform a recalibration of the data set using a quadratic fit.

3.2.3 Statistical Analysis

The acquired images can involve enormous data amounts, as there is a complete spectrum for each pixel. Hence these datasets often comprise file sizes of multiple gigabytes. Therefore, spatial resolution and pixel number, respectively, need to be in a practical balance. Initial, unbiased analysis of the entire imaging dataset can be performed using multiple multivariate statistical analysis (MVA) tools, such as principal component analysis (PCA), maximum autocorrelation factor analysis (MAF) as well as hierarchical cluster analysis. These tools represent excellent approaches for unbiased interrogation of molecular intensity distributions and potential localization to anatomical features of interest. Multiple image analysis strategies have been reported for ToF-SIMS [25, 26] and MALDI IMS data [27]. This top-down approach allows segmentation of a biological sample in distinct regions of interest. This is achieved by detecting (co)variances and correlations in the multivariate data that are encompassed in the MVSA-factors. From the corresponding scores and loadings the variables that are contributing (i.e., mass peak values) to these variances the most can be deduced revealing histology associated chemical changes. This approach facilitates also to outline potential regions of interest (ROI) for subsequent statistical analysis of ROI spectral data and comparative statistics. Alternatively, ROIs can be outlined based on overlay with immunohistochemical stainings subsequently performed on the same or consecutive tissue sections.

3.3 Molecular Identification

While imaging mass spectrometry is a very powerful approach to reveal molecular localization patterns in tissues and cells, there is a need for complementary validation strategies in order to fully exploit the potential of the technique. Initial mass peak annotation is based on accurate mass and reference values reported in the literature or on online depositories. This is typically complemented by subsequent immunohistochemical (IHC) staining on the same or consecutive tissue sections. While this approach allows for verification of the spatial distribution data as well as indicates protein or peptide identity, there are several limitations remaining with respect to chemical specificity and accurate quantification.

3.3.1 Immunohisto chemistry

The most common histochemical validation approaches include immunohistochemistry of the same or consecutive tissue sections using fluorescent microscopy as well as chromogenic detection. Immunohistochemistry is, however, limited by the antibody specificity and difficult to compare in case of truncated protein isoforms leaving preliminary protein annotation in doubt. Moreover, recent studies demonstrated that laser ablation induced distortion effects lead to antigen degradation and increased auto-fluorescence rendering fluorescent immunostainings on the same section to be challenging. Interestingly, chemical imaging with unspecific morphological satins such as H&E and DAPI were not affected by laser ablation induced distortion [28]. This is of great concern as most molecular distributions observed in imaging MS are aligned with histological stains [28].

3.3.2 Mass Spectrometry-Based Protein and Peptide Identification

Due to the limitations of IHC-based approaches with respect to selectivity and specificity as well as quantification further efforts are needed for validating mass peak identities observed in IMS experiments.

In situ peak annotation using mass accuracy is particularly challenging for intact proteins as these experiments are performed in linear mode, where mass accuracy and mass resolution of the TOF instrumentation do not permit identification. Therefore mass spectrometry-based workflows either directly in situ, or on extracts of dissected tissues are the method of choice for unequivocal protein and peptide peak identification [29–31].

The most convenient strategy would therefore involve direct in-situ fragmentation and identification of peptides and proteins. This top-down approach is, however, very challenging for intact protein species due to their size, although significant contributions have been made to advance this strategy using e.g. high-resolution Fourier-transform ion cyclotron resonance (FTICR) instruments. Here, several top-down approaches for in situ protein sequencing have been reported, including in source decay-based protein fragmentation, [32] as well as MS/MS of intact proteins using electron capture dissociation and infrared multiphoton dissociation-based fragmentation for sequence analysis [33]. Another commonly used, straightforward strategy for protein identification in MALDI imaging includes direct on-tissue digestion [29]. Here MALDI IMS is followed by applying trypsin solution onto the same or a consecutive section followed by imaging and fragmentation of the corresponding proteolytic peptides. Identified peptide species are then confirming protein identity by showing similar distribution patterns as the corresponding intact proteins [29]. This elegant, straightforward approach is well established but one has to consider several limitations with respect to lateral diffusion as a consequence of additional liquid deposition during the on-tissue digestion procedure. In addition, only one or two good quality MS/MS spectra can be retrieved from a single spot, since this MALDI-based MS/MS requires higher laser powers resulting in increased sample consumption.

A further convenient strategy to identify unknown protein and peptide peaks is based on tissue extraction. For endogenous peptides (incl. neuropeptides), this extraction is followed by prefractionation using molecular weight cut off (MWCO) filter to avoid convolution by degradation products of larger proteins [34–36]. For neuropeptide analysis a main focus has also been put on limiting postmortem degradation in vivo, ex vivo and in situ using, e.g., heat stabilization and snap freezing [22, 36, 37] prior to sonication and mass cut off filtration enabling large scale neuropetidomics [34–38]. Neuropeptide extracts are then directly analyzed by one- or two-dimensional HPLC hyphenated to electrospray tandem mass spectrometry (LC-ESI-MS/MS) [34–36, 39] (*see* Chapter 14).

In contrast, intact protein extracts require further separation steps using e.g. 1D gel electrophoresis or liquid chromatography using, e.g., a reversed phase (C8) column [31]. This is followed by off line fractionation, fraction collection and automatic MALDI MS interrogation of the collected protein fractions (LC-MALDI). Based on the intact protein MALDI results, protein fractions that contained a protein peak of interest are then subjected to subsequent trypsination followed by identification using LC-ESI-MS/MS [30, 31].

4   MALDI IMS of Proteins and Peptides in Neurodegenerative Diseases

Protein and peptide analysis are of great relevance in neuroscience research and particularly when studying neurodegenerative diseases, where abnormal aggregation and deposition of proteins is a seminal pathological hallmark [40, 41]. For instance, probing neuropeptide and opioid peptide signaling are of essential relevance when investigating Parkinson's disease. Similarly, detection of amyloid beta (Aβ) peptide species is important for studying AD pathology.

Commonly, antibody-based techniques are the dominant method used for in situ detection and quantification of brain peptides. This approach is, however, significantly limited by antibody specificity. This is particularly relevant for opioid peptides that differ in only a few C-terminal amino acids, which compromises the reliability of immunohistochemistry results significantly. Another limitation in antibody-based approaches is the inherent limited throughput, allowing only detection of few species at a time. Furthermore, there are well-known quantification issues since the signal intensity in immunohistochemistry provides only semi quantitative information. LC-MS-based peptide quantification in tissue extracts has been demonstrated to be a powerful approach for characterization of endogenous peptide species [36]. However, dissection and extraction of tissue samples results in loss of spatial information, which highlights the need for techniques that allow in situ analysis. IMS of proteins and peptides has therefore significant advantages over immuno-based techniques particularly with respect to molecular specificity.

An important consideration for imaging of biological tissue samples is the inherent biological variation within the sample groups. This is particularly relevant for IMS of human material. Beyond the inherent patient-to-patient variation, one has to account for technical issues that impact sample quality and increase variation; such as standardized sample collection, sample storage, as well as well-defined and short postmortem delay. This highlights the need for appropriate study design i.e., inclusion of sufficient individuals and matched controls.

An elegant way to avoid the effects of intra-sample group variation is to use single-hemispheric disease models such as unilateral lesion of nigrostratial projections with 6-hydroxy-dopamine (6-OHDA) as a model for Parkinson's disease [42]. These animals develop PD pathology only on one side and the other side can serve as an internal control. This is particularly relevant for mass spectrometry-based studies to account for variation as a result of suppression effects. Using this model, MALDI IMS has been employed to analyze spatial regulations of dynorphin neuropeptides in L-DOPA induced dyskinesia (LID) in experimental PD [22, 43] (Fig. 2a, b). Elevated levels of prodynorphin, the neuropeptide precursor protein of dynorphin peptides, have been previously associated with L-DOPA induced dyskinesia. The lack of specific antibodies permitted accurate annotation of distinct prodynorphin processing products which could therefore not be fully characterized using conventional techniques. MALDI imaging was used to demonstrate that dynorphin B and alpha neoendorphin were significantly elevated in the dorsal lateral striatum in the high dyskinetic group but not for low diskinetic animals. In addition, both dynorphin species correlated positively with LID severity. MALDI imaging has been further successfully applied for protein and peptide imaging in other studies on experimental Parkinsons. Here, IMS was employed to study striatal protein regulstions in 1-methyl-4-phenyl-1,2,3,6-tetrahydropyridine (MPTP) injected rats (Fig. 2b) [44]. Here, MALDI imaging revealed a striatal decrease of the neuronal calmodulin binding protein Pep19, which suggests that altered calcium homeostasis might be associated with neuronal cell death in PD. In a 6-OHDA-based model of experimental PD, ubiquitin, trans-elongation factor 1, hexokinase, and neurofilament M were found down regulated in PD [45].

MALDI Imaging has also been successfully employed to image intact protein species in postmortem spinal cord sections (thoracic) of patients that suffered from amyotrophic lateral sclerosis (ALS) (Fig. 3a) [31]. Here, a C-terminally truncated ubiquitin species (Ubc 1–74, Ubc-T) as well as full length ubiquitin were found to decrease in the ventral horn in ALS suggesting an altered protein turnover in ALS (Fig. 3b). These findings are of particular relevance, since in situ differentiation of C-terminal sequence truncations cannot be achieved with antibody-based techniques due to limited specificity. Therefore, these data highlight the potential of imaging MS for studies on intact peptide and protein regulations in situ.

Along with neuropeptide and small protein analysis in neurodegenerative diseases, MALDI IMS has been demonstrated to be a very powerful technique to probe amyloid pathology in Alzheimer's disease [7, 46–52]. Here, Aβ truncations in individual plaques have been characterized in APP23 transgenic mice (Fig. 4a) [7, 46]. The authors reported a significantly higher content of Aβ 1–40 than Aβ 1–42. Similarly, our group reported a comprehensive study on profiling brain-wide Aβ profiles in transgenic animals carrying the Swedish and Arctic mutation in the *APP* gene (tgArcSwe) (Fig. 4b) [47]. In this study, multivariate image analysis was employed to outline pathological features that resemble Aβ plaques (Fig. 4b-I). Inspection of the corresponding loadings revealed distinct Aβ peptide species to be the most prominent variables causing the chemical differences obtained by cluster analysis (Fig. 4b-II–IV).

In another study on triple-transgenic animal model of AD (3xtg), MALDI IMS revealed an AD pathology-associated decrease of the synaptic protein neurogranin (NGRN), which is in line with clinical findings where CSF levels of NGRN were demonstrated to be a potential biomarker reflecting AD pathology associated synaptic degeneration [48].

Recently, MALDI IMS was employed for human AD tissue, where amyloid peptide profiles of individual plaques were delineated [49]. Here, Aβ1–42 and Aβ1–43 were found to be selec-

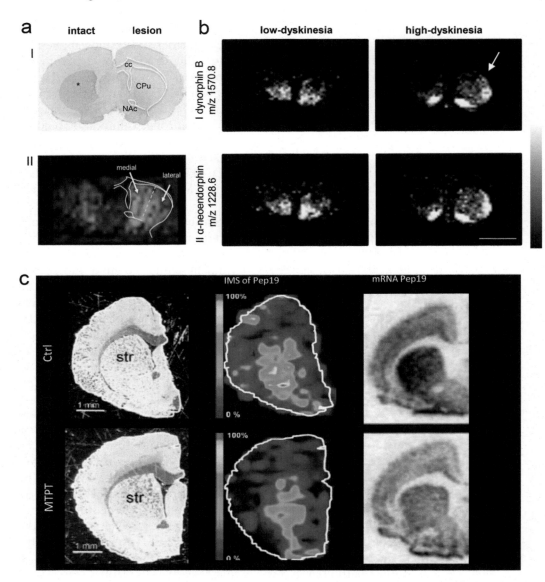

Fig. 2 Imaging mass spectrometry in Parkinson's disease. Neuropeptides in L-DOPA-induced dyskinesia. (**a**) I, Unilateral 6-OHDA injection leads to dopamine depletion (as illustrated by tyrosine hydroxylase, TH-immunostaining *). L-DOPA therapy results in two distinct groups with low and high dyskinesia. II, Striatal sections were analyzed by MALDI IMS and regions of interest (ROI) were assigned for spectral data extraction. Scalebar: 2 mm (**b**) Dynorphin peptides, I dynorphin B and II alpha neorndorphin, displayed a significant ($p > 0.05$) relative increase in the dorsolateral striatum of high- dyskinetic (HD) compared to low-dyskinetic (LD) animals and lesion controls (LC) (**c**) MALDI IMS reveals decrease of striatal Pep19 levels in MPTP lesioned rats, as verified by in situ hybridization and LCMS in tissue extracts

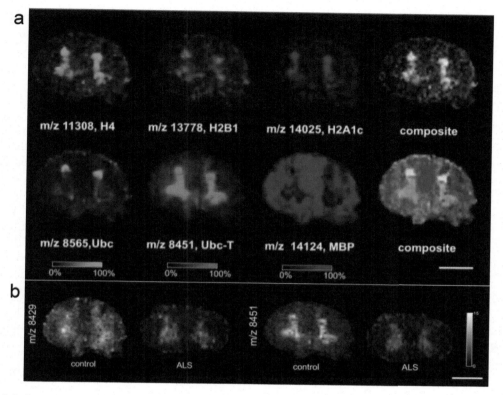

Fig. 3 Delineating protein dynamics in ALS using imaging mass spectrometry. (**a**) Imaging MS of spinal cord proteins reveals characteristic protein distribution patterns that are well in line with anatomical regions. This includes the gray matter (histones, truncated ubiquitin 1–74 (Ubc-T)), the dorsal horn of the gray matter (ubiquitin, Ubc), and white matter (myelin basic protein, MBP). (**b**) Multivariate statistics reveal significant decrease for two protein peaks, including Ubc-T (*m/z* 8451) in the gray matter of ALS patients compared to controls. Scalebar: 2 mm

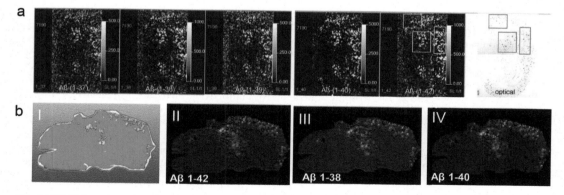

Fig. 4 MALDI imaging of amyloid peptides in mouse models of Alzheimer's disease. (**a**) MALDI-based neuropeptide imaging in APP23 mice illustrates colocalization of various truncated Aβ-peptide to plaques. (**b**) MALDI IMS in 18–month-old tgAPPArcSwe mice. (**b**) I, Image analysis using hierarchical cluster analysis (HCA, bisecting k-means) delineates histological features resembling plaque pathology (yellow, green). (**b**) II–IV, Inspection of the corresponding variables in the clusters that cause this difference reveals major Aβ species

tively deposited in senile plaques while other C-terminally truncated Aβ species including, Aβ1–36, 1–37, 1–38, 1–39, 1–40, and Aβ1–41 were primarily detected in leptomeningeal blood vessels.

5 Conclusions

The developments and applications of MS imaging presented here highlight the immense potential of the technique for studying spatial protein dynamics in complex biological tissues. Major challenges associated with sample preparation, sample throughput and protein identification are more and more being overcome by recent developments. Moreover, standardized protocols and data formats significantly facilitate IMS-based applications in clinical settings. In summary, MALDI imaging is increasingly becoming a valued alternative for protein and peptide imaging and has proved to show great potential for delineating molecular mechanisms underlying neurodegenerative disease pathology.

Acknowledgements

Funding was provided by The Swedish Research Council VR (no. 2014-6447 and no.2018-02181, J.H.; no. 2013-2546, H.Z.; no. 2017-00915, K.B.), Alzheimerfonden (J.H., K.B.), Hjärnfonden (K.B.), Åke Wiberg Stiftelse (J.H.), Ahlén Stiftelsen (J.H.), Stiftelsen Gamla Tjänarinnor (J.H., H.Z., K.B.), Torsten Söderberg Foundation (K.B.), the Knut and Alice Wallenberg Foundation (H.Z.), and the European Research Council (681712, H.Z.).

References

1. Karas M, Hillenkamp F (1988) Laser desorption ionization of proteins with molecular masses exceeding 10000 Daltons. Anal Chem 60(20):2299–2301

2. Fenn JB, Mann M, Meng CK, Wong SF, Whitehouse CM (1989) Electrospray ionization for mass-spectrometry of large biomolecules. Science 246(4926):64–71

3. McDonnell LA, Heeren RMA (2007) Imaging mass spectrometry. Mass Spectrom Rev 26(4):606–643. https://doi.org/10.1002/mas.20124

4. Cornett DS, Reyzer ML, Chaurand P, Caprioli RM (2007) MALDI imaging mass spectrometry: molecular snapshots of biochemical systems. Nat Methods 4(10):828–833. https://doi.org/10.1038/nmeth1094

5. Hanrieder J, Malmberg P, Ewing AG (2015) Spatial neuroproteomics using imaging mass spectrometry. Biochim Biophys Acta 1854(7):718–731. https://doi.org/10.1016/j.bbapap.2014.12.026

6. Hanrieder J, Phan NT, Kurczy ME, Ewing AG (2013) Imaging mass spectrometry in neuroscience. ACS Chem Neurosci 4(5):666–679. https://doi.org/10.1021/cn400053c

7. Seeley EH, Caprioli RM (2008) Molecular imaging of proteins in tissues by mass spectrometry. Proc Natl Acad Sci U S A 105(47):18126–18131. https://doi.org/10.1073/pnas.0801374105

8. Vickerman JC (2011) Molecular imaging and depth profiling by mass spectrometry--SIMS, MALDI or DESI? Analyst 136(11):2199–2217. https://doi.org/10.1039/c1an00008j

9. Takats Z, Wiseman JM, Gologan B, Cooks RG (2004) Mass spectrometry sampling under ambient conditions with desorption electrospray ionization. Science 306(5695):471–473. https://doi.org/10.1126/science.1104404

10. Caprioli RM, Farmer TB, Gile J (1997) Molecular imaging of biological samples: localization of peptides and proteins using MALDI-TOF MS. Anal Chem 69(23):4751–4760

11. Gustafsson JO, Oehler MK, McColl SR, Hoffmann P (2010) Citric acid antigen retrieval (CAAR) for tryptic peptide imaging directly on archived formalin-fixed paraffin-embedded tissue. J Proteome Res 9(9):4315–4328. https://doi.org/10.1021/pr9011766

12. Meding S, Martin K, Gustafsson OJ, Eddes JS, Hack S, Oehler MK, Hoffmann P (2013) Tryptic peptide reference data sets for MALDI imaging mass spectrometry on formalin-fixed ovarian cancer tissues. J Proteome Res 12(1):308–315. https://doi.org/10.1021/pr300996x

13. Goodwin RJA, Dungworth JC, Cobb SR, Pitt AR (2008) Time-dependent evolution of tissue markers by MALDI-MS imaging. Proteomics 8(18):3801–3808. https://doi.org/10.1002/pmic.200800201

14. Goodwin RJA, Lang AM, Allingham H, Boren M, Pitt AR (2010) Stopping the clock on proteomic degradation by heat treatment at the point of tissue excision. Proteomics 10(9):1751–1761. https://doi.org/10.1002/pmic.200900641

15. Goodwin RJA, Nilsson A, Borg D, Langridge-Smith PRR, Harrison DJ, Mackay CL, Iverson SL, Andren PE (2012) Conductive carbon tape used for support and mounting of both whole animal and fragile heat-treated tissue sections for MALDI MS imaging and quantitation. J Proteomics 75(16):4912–4920. https://doi.org/10.1016/j.jprot.2012.07.006

16. Hanrieder J, Ljungdahl A, Andersson M (2012) MALDI imaging mass spectrometry of neuropeptides in Parkinson's disease. J Vis Exp (60). https://doi.org/10.3791/3445

17. Shariatgorji M, Kallback P, Gustavsson L, Schintu N, Svenningsson P, Goodwin RJA, Andren PE (2012) Controlled-pH tissue cleanup protocol for signal enhancement of small molecule drugs analyzed by MALDI-MS imaging. Anal Chem 84(10):4603–4607. https://doi.org/10.1021/ac203322q

18. Seeley EH, Oppenheimer SR, Mi D, Chaurand P, Caprioli RM (2008) Enhancement of protein sensitivity for MALDI imaging mass spectrometry after chemical treatment of tissue sections. J Am Soc Mass Spectrom 19(8):1069–1077. https://doi.org/10.1016/j.jasms.2008.03.016

19. Martin-Lorenzo M, Balluff B, Sanz-Maroto A, van Zeijl RJ, Vivanco F, Alvarez-Llamas G, McDonnell LA (2014) 30mum spatial resolution protein MALDI MSI: in-depth comparison of five sample preparation protocols applied to human healthy and atherosclerotic arteries. J Proteomics 108:465–468. https://doi.org/10.1016/j.jprot.2014.06.013

20. Stoeckli M, Staab D, Wetzel M, Brechbuehl M (2014) iMatrixSpray: a free and open source sample preparation device for mass spectrometric imaging. Chimia (Aarau) 68(3):146–149. https://doi.org/10.2533/chimia.2014.146

21. Aerni HR, Cornett DS, Caprioli RM (2006) Automated acoustic matrix deposition for MALDI sample preparation. Anal Chem 78(3):827–834. https://doi.org/10.1021/ac051534r

22. Hanrieder J, Ljungdahl A, Falth M, Mammo SE, Bergquist J, Andersson M (2011) L-DOPA-induced dyskinesia is associated with regional increase of striatal dynorphin peptides as elucidated by imaging mass spectrometry. Mol Cell Proteomics 10(10):M111.009308. https://doi.org/10.1074/mcp.M111.009308

23. Yang J, Caprioli RM (2011) Matrix sublimation/recrystallization for imaging proteins by mass spectrometry at high spatial resolution. Anal Chem 83(14):5728–5734. https://doi.org/10.1021/ac200998a

24. Puolitaival SM, Burnum KE, Cornett DS, Caprioli RM (2008) Solvent-free matrix dry-coating for MALDI imaging of phospholipids. J Am Soc Mass Spectrom 19(6):882–886. https://doi.org/10.1016/j.jasms.2008.02.013

25. Graham DJ, Castner DG (2012) Multivariate analysis of ToF-SIMS data from multicomponent systems: the why, when, and how. Biointerphases 7(1–4):49. https://doi.org/10.1007/s13758-012-0049-3

26. Henderson A, Fletcher JS, Vickerman JC (2009) A comparison of PCA and MAF for ToF-SIMS image interpretation. Surf Interface Anal 41(8):666–674. https://doi.org/10.1002/sia.3084

27. Deininger S-O, Ebert MP, Fuetterer A, Gerhard M, Roecken C (2008) MALDI imaging combined with hierarchical clustering as a new tool for the interpretation of complex human cancers. J Proteome Res 7(12):5230–5236. https://doi.org/10.1021/pr8005777

28. Kaya I, Michno W, Brinet D, Iacone Y, Zanni G, Blennow K, Zetterberg H, Hanrieder

J (2017) Histology-compatible MALDI mass spectrometry based imaging of neuronal lipids for subsequent immunofluorescent staining. Anal Chem 89(8):4685–4694. https://doi.org/10.1021/acs.analchem.7b00313

29. Groseclose MR, Andersson M, Hardesty WM, Caprioli RM (2007) Identification of proteins directly from tissue: in situ tryptic digestions coupled with imaging mass spectrometry. J Mass Spectrom 42(2):254–262. https://doi.org/10.1002/jms.1177

30. Andersson M, Groseclose MR, Deutch AY, Caprioli RM (2008) Imaging mass spectrometry of proteins and peptides: 3D volume reconstruction. Nat Methods 5(1):101–108. https://doi.org/10.1038/nmeth1145

31. Hanrieder J, Ekegren T, Andersson M, Bergquist J (2012) MALDI imaging mass spectrometry of human post mortem spinal cord in amyotrophic lateral sclerosis. J Neurochem 124:695–707. https://doi.org/10.1111/jnc.12019

32. Debois D, Bertrand V, Quinton L, De Pauw-Gillet MC, De Pauw E (2010) MALDI-in source decay applied to mass spectrometry imaging: a new tool for protein identification. Anal Chem 82(10):4036–4045. https://doi.org/10.1021/ac902875q

33. Kiss A, Smith DF, Reschke BR, Powell MJ, Heeren RM (2014) Top-down mass spectrometry imaging of intact proteins by laser ablation ESI FT-ICR MS. Proteomics 14(10):1283–1289. https://doi.org/10.1002/pmic.201300306

34. Skold K, Svensson M, Kaplan A, Bjorkesten L, Astrom J, Andren PE (2002) A neuroproteomic approach to targeting neuropeptides in the brain. Proteomics 2(4):447–454. https://doi.org/10.1002/1615-9861(200204)2:4<447::aid-prot447>3.0.co;2-a

35. Svensson M, Skold K, Svenningsson P, Andren PE (2003) Peptidomics-based discovery of novel neuropeptides. J Proteome Res 2(2):213–219

36. Svensson M, Skold K, Nilsson A, Falth M, Svenningsson P, Andren PE (2007) Neuropeptidomics: expanding proteomics downwards. Biochem Soc Trans 35(Pt 3):588–593. https://doi.org/10.1042/bst0350588

37. Che FY, Lim J, Pan H, Biswas R, Fricker LD (2005) Quantitative neuropeptidomics of microwave-irradiated mouse brain and pituitary. Mol Cell Proteomics 4(9):1391–1405. https://doi.org/10.1074/mcp.T500010-MCP200

38. Fricker LD (2007) Neuropeptidomics to study peptide processing in animal models of obesity.

Endocrinology 148(9):4185–4190. https://doi.org/10.1210/en.2007-0123

39. Yin P, Hou X, Romanova EV, Sweedler JV (2011) Neuropeptidomics: mass spectrometry-based qualitative and quantitative analysis. Methods Mol Biol 789:223–236. https://doi.org/10.1007/978-1-61779-310-3_14

40. Blennow K, de Leon MJ, Zetterberg H (2006) Alzheimer's disease. Lancet 368(9533):387–403. https://doi.org/10.1016/s0140-6736(06)69113-7

41. Jucker M, Walker LC (2013) Self-propagation of pathogenic protein aggregates in neurodegenerative diseases. Nature 501(7465):45–51. https://doi.org/10.1038/nature12481

42. Ungerstedt U (1968) 6-Hydroxy-dopamine induced degeneration of central monoamine neurons. Eur J Pharmacol 5(1):107

43. Ljungdahl A, Hanrieder J, Faelth M, Bergquist J, Andersson M (2011) Imaging mass spectrometry reveals elevated nigral levels of dynorphin neuropeptides in L-DOPA-induced dyskinesia in rat model of Parkinson's disease. PLoS One 6(9):e25653. https://doi.org/10.1371/journal.pone.0025653

44. Skold K, Svensson M, Nilsson A, Zhang XQ, Nydahl K, Caprioli RM, Svenningsson P, Andren PE (2006) Decreased striatal levels of PEP-19 following MPTP lesion in the mouse. J Proteome Res 5(2):262–269. https://doi.org/10.1021/pr050281f

45. Stauber J, Lemaire R, Franck J, Bonnel D, Croix D, Day R, Wisztorski M, Fournier I, Salzet M (2008) MALDI imaging of formalin-fixed paraffin-embedded tissues: application to model animals of Parkinson disease for biomarker hunting. J Proteome Res 7(3):969–978. https://doi.org/10.1021/pr070464x

46. Stoeckli M, Staab D, Staufenbiel M, Wiederhold KH, Signor L (2002) Molecular imaging of amyloid beta peptides in mouse brain sections using mass spectrometry. Anal Biochem 311(1):33–39

47. Carlred L, Michno W, Kaya I, Sjovall P, Syvanen S, Hanrieder J (2016) Probing amyloid-beta pathology in transgenic Alzheimer's disease (tgArcSwe) mice using MALDI imaging mass spectrometry. J Neurochem 138(3):469–478. https://doi.org/10.1111/jnc.13645

48. Kvartsberg H, Duits FH, Ingelsson M, Andreasen N, Ohrfelt A, Andersson K, Brinkmalm G, Lannfelt L, Minthon L, Hansson O, Andreasson U, Teunissen CE, Scheltens P, Van der Flier WM, Zetterberg H, Portelius E, Blennow K (2015) Cerebrospinal fluid levels of the synaptic protein neurogranin correlates with cognitive decline in prodromal Alzheimer's

disease. Alzheimers Dement 11(10):1180–1190. https://doi.org/10.1016/j.jalz.2014.10.009

49. Kakuda N, Miyasaka T, Iwasaki N, Nirasawa T, Wada-Kakuda S, Takahashi-Fujigasaki J, Murayama S, Ihara Y, Ikegawa M (2017) Distinct deposition of amyloid-beta species in brains with Alzheimer's disease pathology visualized with MALDI imaging mass spectrometry. Acta Neuropathol Commun 5(1):73. https://doi.org/10.1186/s40478-017-0477-x

50. Kaya I, Brinet D, Michno W, Başkurt M, Zetterberg H, Blenow K, Hanrieder J (2017) Novel Trimodal MALDI imaging mass spectrometry (IMS3) at 10 μm reveals patial lipid and peptide correlates implicated in Aβ plaque pathology in Alzheimer's disease. ACS Chem Neurosci 8(12):2778–2790. https://doi.org/10.1021/acschemneuro.7b00314

51. Kaya I, Zetterberg H, Blennow K, Hanrieder J (2018) Shedding light on the molecular pathology of amyloid plaques in transgenic Alzheimer's disease mice using multimodal MALDI imaging mass spectrometry. ACS Chem Neurosci 18;9(7):1802–1817. https://doi.org/10.1021/acschemneuro.8b00121

52. Michno W, Nyström S, Wehrli P, Lashley T, Brinkmalm G, Guerard L, Syvänen S, Sehlin D, Kaya I, Brinet D, Nilsson KPR, Hammarström P, Blennow K, Zetterberg H, anrieder J (2019) Pyroglutamation of amyloid-βx-42 (Aβx-42) followed by Aβ1-40 deposition underlies plaque polymorphism in progressing Alzheimer's disease pathology. J Biol Chem 294(17):6719–6732. https://doi.org/10.1074/jbc.RA118.006604

## Chapter 16

A Combined Cellomics and Proteomics Approach to Uncover Neuronal Pathways to Psychiatric Disorder

Martina Rosato, Titia Gebuis, Iryna Paliukhovich, Sven Stringer, Patrick F. Sullivan, August B. Smit, and Ronald E. van Kesteren

Abstract

Studying biological mechanisms underlying neuropsychiatric disorders is highly challenging as many risk genes are associated with these disorders. This complexity requires research approaches to reliably dissect the cell biology of the risk genes involved. Here, we describe a combined cellomics–proteomics approach that allows (a) medium-throughput functional screening and unbiased selection of important risk genes, and (b) discovery of common functional pathways and interactome connections of selected risk genes. The overlay of pathway and proteome data from selected genes in a biological context can be used to formulate new testable hypothesis of both the genetics and the biology of the disorders.

Key words Cellomics, High-content screening, Proteomics, Psychiatric disorders

1 Introduction

In the last decade, large-scale genetic analyses have shown that the biology of human brain diseases is highly complex and involves many genes. This complexity becomes particularly challenging when studying neuropsychiatric disorders involving a multitude of different biological processes, such as the developmental and specification of neuronal cell types, neuronal circuitry formation and maintenance, and the formation and function of synaptic connections. Many neuropsychiatric disorders have high heritability (e.g., 0.80 in autism, 0.75 in bipolar disorder, and 0.81 in schizophrenia) and major efforts have proven successful in uncovering loci and genes involved [1]. Schizophrenia (SCZ) in particular was shown to be highly polygenic with many common and some rare genetic variants contributing to its pathogenesis [2]. Recent common variant genome-wide association studies (GWAS) highlighted up to 145 genetic loci that are significantly associated with SCZ, together encompassing about a thousand genes [3, 4]. Furthermore, smaller

Ka Wan Li (ed.), *Neuroproteomics*, Neuromethods, vol. 146,
https://doi.org/10.1007/978-1-4939-9662-9_16, © Springer Science+Business Media, LLC, part of Springer Nature 2019

numbers of rare copy number variants (CNVs), often impacting many genes, were also found to be associated with SCZ [5–9].

Studies towards revealing causes of complex disorders such as SCZ, or designing potential treatments, therefore require systematic dissection of the biological functions of many genes and the robust identification of shared functional pathways and interactions between them. Indeed, there is a need for methods that (a) permit large-scale functional screening in cellular models or model organisms that are relevant to the disease, (b) allow subsequent unbiased target selection, and (c) enable adequate follow-up studies such as proteomics to uncover pathways and functional interactions underlying the disease. Here, we describe the integrated use of two powerful technologies—cellomics and proteomics—and how these can be applied to cultured neurons to systematically uncover the functions of psychiatric disease risk genes in neuronal network development and contribute to the functional interpretation of the increasing amount of available genetics data.

Cellomics or cellular high-content screening (HCS) is a powerful microscopy-based method to collect in-depth data on morphological and cell biological changes, for instance during neuronal development [10–15]. A main advantage is that image acquisition and image analyses are fully automated on HCS instruments, thus delivering unbiased large-scale data sets. The use of multiwell plates (usually 96 or 384 wells) further permits the simultaneous analysis of many experimental conditions. Finally, HCS can assess an entire population of neurons per culture well instead of just a few cells that are manually selected, and thus adequately capture population characteristics in addition to single-cell properties. Together, these features of HCS increase the statistical power to detect experimentally induced alterations in cellular morphology, and the reliability and reproducibility of the morphological phenotypes detected. Another advantage of HCS is that it provides information on many different cellular parameters in parallel. The combination of multiple cellular stainings permits quantification of many aspects of cellular biology simultaneously, i.e., it is high-content, and allows hypothesis-free research designs that could lead to the discovery of unexpected phenotypes. Here we describe HCS with RNA interference (RNAi) to study the role of risk genes for psychiatric diseases in neuronal development, and to determine common neuronal phenotypes for groups of genes.

Behind the cell biological phenotypes captured by HCS are mutations that alter or abolish the expression of risk genes. These will in many cases directly or indirectly change the expression of other genes and proteins as well. Proteomics analysis, i.e., the large-scale analysis of protein identity and quantity of tissues, cells, or subcellular compartments, is a powerful method to systematically analyze protein dysregulation resulting from genetic modification or experimental RNAi. The large-scale analysis of dysregulated proteins following the experimental perturbation of risk genes serves different purposes. Firstly, when multiple short-hairpin RNAs (shRNAs) are used to knockdown a gene of interest, the resulting alterations in protein expression can be compared in order to separate on-target from off-target effects. Robust and reproducible changes in protein expression can then be used to probe the functional protein interactome and pathway analyses can be performed on a per gene basis to identify common pathways or final regulators in these pathways. Finally, overlapping pathway or interactome data of multiple genes can be used to identify causal risk genes and regulatory modules hidden in existing GWAS data and formulate new testable hypotheses of the development of complex neuropsychiatric disorders.

In this chapter, we describe a high-throughput workflow that combines cellomics and proteomics for the study of biological processes underlying complex polygenic disorders (Fig. 1). We showcase an analysis pipeline for HCS data that maximizes the extraction of reliable and significant hits, and we provide suggestions as to how systematic proteomic analyses of multiple RNAi interventions may lead to the identification of robust pathway and interactome maps that facilitate the discovery of additional risk genes.

2   Data Acquisition

2.1   Neuronal Cultures

Primary hippocampal or cortical neuronal cultures are prepared from wildtype C57Bl/6 embryonic day 18 (E18) mouse embryos. Tissue is collected in Hanks balanced salts solution (Sigma, St. Louis, MO), buffered with 7 mM HEPES (Gibco, Waltham, MA), and incubated for 25 min in HBBS (Gibco) containing 0.25% trypsin (Gibco) at 37 °C. After washing, neurons are triturated using fire-polished Pasteur pipettes, counted, and plated in Neurobasal medium supplemented with 2% B-27, 2% HEPES (pH = 7.4), 0.25% glutamine and 1% Pen Strep (all from Gibco). Cells are plated in multiwell plates that were previously coated with poly-D-lysine (Sigma). Neurons are plated at different densities, i.e., 12.5 K cells/well in 96-well glass-bottom plates (Greiner Bio-One, Kremsmünster, Austria) for morphological analyses and 300 K cells/well in 12-well plates (Greiner Bio-One) for protein extraction, and cultured at 37 °C/5% CO_2 for 7, 14 or 21 days. In principle, these experiments can also be performed with human iPSC-derived neurons in order to capture the total genetic complexity underlying a particular disorder. Preparing high-quality neuronal cultures is critical and several precautions are required to reduce technical variation. For example, we only seed neurons in the inner 60 wells of a 96-well plate while keeping the outer wells filled with medium. This reduces edge effects due to the higher evaporation of the medium from the outer wells over the total duration of an experiment.

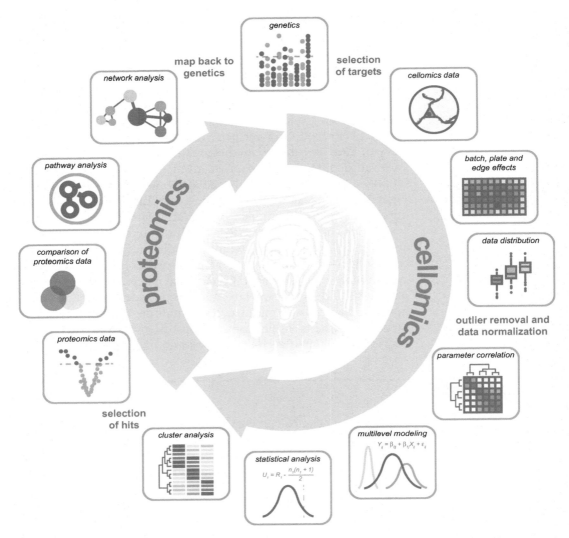

Fig. 1 Overview of the experimental approach. A schematic representation of the combined cellomics and proteomics analysis of risk genes for neuropsychiatric disorders showing critical steps and decision moments in the procedure. Details for every step are provided in the main text

2.2 RNA Interference   Short-hairpin RNA (shRNA) constructs are ordered as bacterial glycerol stocks (MISSION library, Sigma) and grown on agar plates with LB medium and 1% ampicillin. Single colonies are picked and expanded for DNA extraction using QIAprep spin mini columns (Qiagen, Hilden, Germany). HEK 293 T cells are transfected with the plasmid DNA together with envelope and packaging constructs. One day after transfection, medium is replaced with Optimem medium (Gibco) completed with 1% Pen/Strep and 1% glutamine. On day 3, the medium is collected and centrifuged at $1000 \times g$ for 5 min; the supernatant containing the viral particles is filtered (0.45 µm pore size) and aliquoted. To test for transduction

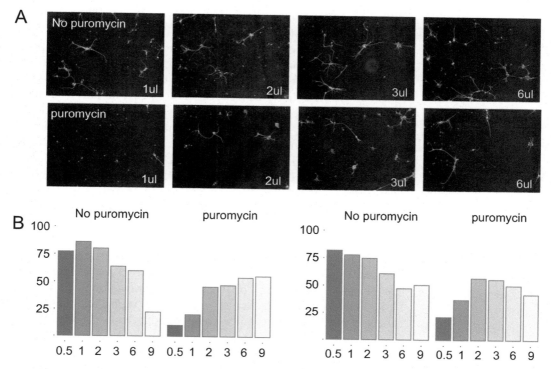

Fig. 2 Determining lentiviral transduction efficiencies. (**a**) Representative images of primary neuron cultures with increasing concentrations of virus showing effects on cell viability in the absence or presence of puromycin. (**b**) Virus transduction efficiency is determined as the concentration of virus that produces maximal protection against puromycin treatment and at the same time minimal cell loss in the absence of puromycin. In these examples the optimal amount of virus would be 6 μl and 2 μl, respectively

efficacy, neurons are plated and infected at day 1 in vitro (DIV1) with 0.5, 1, 3, 6, or 9 μl virus in 200 μl culture medium per well. At DIV2, puromycin (0.2 mg/ml, Gibco) is added and cells are fixed at DIV7 and stained with Hoechst (1:10,000; Invitrogen, Carlsbad, CA) and anti-MAP 2 (1:5000; Bio-connect, Huissen, The Netherlands). Infection efficiency is determined as the percentage of living cells compared to untreated noninfected, control wells. Based on these results, optimal virus concentrations are determined for shRNA screening (Fig. 2).

Proper characterization of shRNAs is critical. Every shRNA spans about 20–22 nucleotides of the target mRNA, and although mRNA specificity is largely determined by a six-nucleotide seed sequence within the shRNA, the chances of off-target effects are considerable, in particular because most of these shRNAs have not been tested in primary neurons in a high-content setting. In addition, different shRNAs targeting the same gene may produce different levels of knockdown and therefore also different phenotypes. We normally use 5 different shRNAs per gene. Comparing the phenotypic effects of these 5 shRNA and searching for common effects among a minimum subset of 3 of them will result in the most reliable phenotype per gene.

The timing of shRNA infection depends on the specific research hypothesis. Hippocampal primary cultures develop over several weeks. Until DIV2, cultures are very immature and no electrical network activity can be observed [16], at DIV3 an active network starts to emerge and becomes synchronized around DIV7, and the network further matures until DIV14 [16]. To study effects on neuronal development, we normally perform shRNA infections at DIV1–2 and fix and stain cultures for HCS at multiple time points, i.e., DIV7, DIV14 and DIV21 (Fig. 3). However, if the experiment is focused on an acute impairment of the network, virus infection, cell fixation and staining can also be performed on the mature network, e.g., at DIV14–21.

2.3 Experimental Design

Even small-scale RNAi screens require many neuronal cultures to be analyzed. For neurodevelopmental effects we perform end-point measurements at 3 different time points (DIV7, DIV14 and DIV21). With 5 shRNAs per gene and 3 replicates per shRNA, analysis of 100 genes require 4500 independent cell cultures, and a minimum of 75 × 96-well plates, excluding controls. We normally process such a screen in batches of 6 or 9 plates. On each plate, we put medium in the outer 36 wells, infect neurons with different shRNAs in 45 wells, and the remaining 15 wells are used for negative and positive controls. Negative controls include a scrambled shRNA (a random shRNA sequence not targeting any gene in the cell type used) and untreated (noninfected) cultures; the positive control is an shRNA with a known effect on the biological process of interest. All controls are always present on every plate with a sufficient number of replicates, preferably 5 per plate, to allow accurate, outlier-insensitive, within-plate normalization of HCS data. For experimental shRNAs we use 3 independent replicates divided over different culture plates to minimize false positives and false negatives due to plate effects. To minimize plate position effects, the order of treatments (controls and experimental shRNAs) on each plate is randomized.

2.4 Cellomics

To visualize the effects of RNAi on neuronal network development, neuronal cultures are fixed at DIV7, DIV14 or DIV21 and stained with markers for dendrite growth and synaptogenesis. Dendrite growth is visualized with antibodies against MAP 2 or β-tubulin, while monitoring synaptogenesis requires the use of both presynaptic (e.g., anti-synapsin or anti-VAMP) and postsynaptic (anti-PSD-95 or anti-homer) markers. Nuclear staining (e.g., Hoechst or DAPI) is used to identify cells. Using appropriate fluorescently labeled secondary antibodies, dendritic and synaptic markers are visualized in different channels, while UV illumination is used to visualize nuclei.

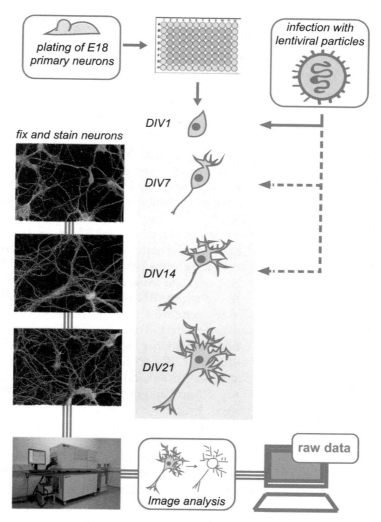

Fig. 3 Cellomics experimental design. Primary neurons from embryonic day 18 (E18) mouse pups are isolated and seeded into the inner wells of 96-well cell culture plates, keeping the outer wells filled with medium only. Neurons are infected with shRNA-encoding lentiviral particles at day 1 in vitro (DIV1), or alternatively, depending on the research question, on DIV7 or DIV14. Cultures are fixed and stained at the desired time points and images are automatically acquired using a high-content microscopy instrument. Images are then analyzed in an automated manner using high-content image analysis software. Multiparametric data is collected for further explorative and statistical analysis

Automated image acquisition and analysis are performed using an integrated HCS microscopy system. We use a CellInsight CX7 (ThermoFischer, Waltham, MA) or Opera LX (PerkinElmer, Waltham, MA) HCS platform in combination with the Columbus image data storage and analysis system (PerkinElmer, Waltham, MA). Typically, ~1000 cells are imaged per well at 10× magnification, and neuron numbers, dendrite numbers, dendrite lengths, and dendrite morphology are determined based on MAP 2 or β-tubulin staining. Next, the same culture wells are imaged at 40× magnification and synapse numbers and morphology are determined based on the combined dendritic and synaptic stainings. Image analysis typically follows the following steps (Fig. 4): (a) detection of nuclei, (b) identification of neurons based on MAP 2 staining and/or neuronal morphology, (c) detection of dendrites originating from neurons followed by extraction of various dendritic parameters, (d) detection of pre- and postsynaptic puncta close to dendrites followed by extraction of various pre-and postsynaptic parameters, and (e) determining the number of co-localized and non-co-localized pre- and postsynaptic puncta.

2.5   Proteomics

To study the effects of RNAi on the overall protein content of neurons, we plate neurons in 12 well plates at a seeding density of 300 K neurons per well and subjected RNAi followed by protein isolation and mass spectrometry (MS) (Fig. 5). Neurons are transfected with shRNAs on DIV1. We usually include all 5 shRNAs for every gene we study, independent of whether or not a phenotype was detected. This allows the detection of protein changes with high confidence while distinguishing on-target from off-target effects at the molecular level. As negative controls, scrambled shRNA-treated and untreated samples are used. For all conditions, i.e., control and experimental, at least 3 independent replicates are used. Proteins are extracted from the cultures when the first phenotypic alterations are observed. We collect protein samples as early as possible, before a strong morphological phenotype is observed, to increase the chance of detecting protein changes that are a direct effect of the knockdown and not the end result of the morphological change. For protein extraction we first wash out the medium with PBS (Gibco) to avoid the inclusion of medium proteins, then gently scrape cells in PBS complemented with protease inhibitor (Roche, Bazel, Switzerland), spin down samples at 3000 × *g* for 5 min at 4 °C and dissolve the pellet in Laemmli buffer. Samples are separated on SDS-PAGE and prepared for MS analysis as described previously [17–19] (*see* Chapters 4, 8 and 11 for protocols). In brief, the SDS-PAGE gel lanes are cut in smaller pieces, destained with two cycles of incubation with 50 nM ammonium bicarbonate (Fluka, Steinheim, Germany)/50% acetonitrile (JT Baker, Deventer, The Netherlands) and dried with a solution of 100% acetonitrile. The gel fragments are then incubated in trypsin solution (Promega, Madison, WI) overnight at 37 °C. Digested peptides are extracted with two cycles of incubation with 0.1% trifluoroacetic acid (Applied Biosystems, Warrington, UK)/50% acetonitrile and one incubation step with 0.1% trifluoroacetic acid/80% acetonitrile. The peptide solution is dried in a speedvac, dissolved in 0.1% acetic acid solution and loaded into an Ultimate 3000 LC system (Dionex, Thermo Scientific, Whaltam, MA) and

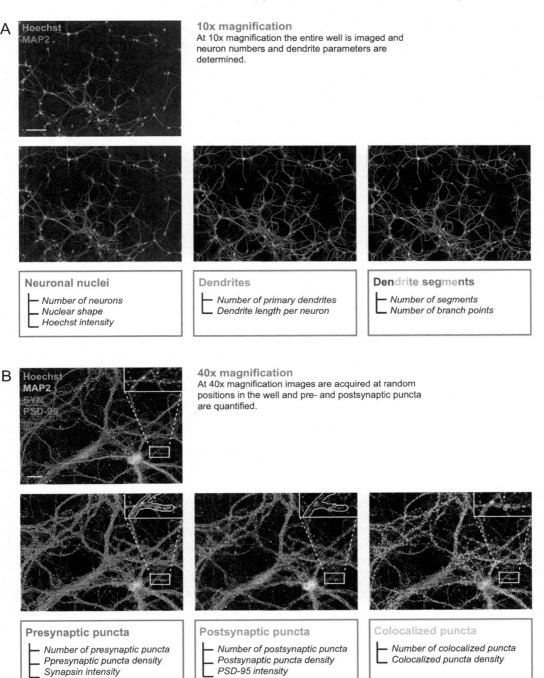

Fig. 4 High-content image analysis. (**a**) Images acquired at 10× magnification are used to determine neuron numbers, nuclear parameters, and dendrite parameters based on Hoechst and MAP 2 staining. (**b**) Images acquired at 40× magnification are used to determine synaptic parameters based on Hoechst, MAP 2, synapsin, and PSD-95 staining

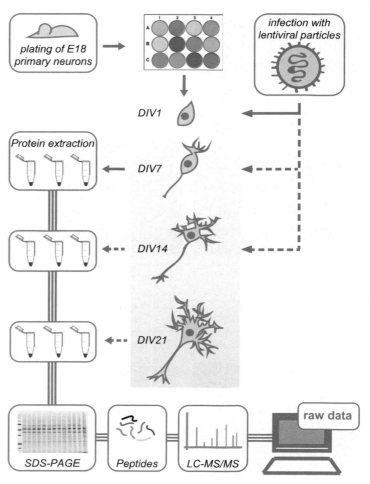

Fig. 5 Proteomics experimental design. Primary neurons from embryonic day 18 (E18) mouse pups are isolated and seeded into 12-well cell culture plates. Neurons are infected with shRNA-encoding lentiviral particles at day 1 in vitro (DIV1), or alternatively, depending on the research question, on DIV7 or DIV14. Cultures are lysed and protein is extracted at the desired time points and run on SDS-PAGE. The gel lanes are then cut in pieces and in-gel tryptic digestion is performed. The resulting peptide samples are analyzed by LC-MS/MS

then on a TripleTop 5600 MS system (Sciex, Framingham, MA) (for further details, *see also* Chapter 11).

3 Data Analysis

To maximize the extraction of disease-relevant data from our combined cellomics and proteomics results, we designed an analysis pipeline consisting of several steps of data quality check, data scaling/normalization, and statistical analysis (Fig. 1).

3.1   Quality Check of Cellomics Data

Cellomics experiments yield multiparametric data sets. Depending on the number of fluorescent channels and how image analysis algorithms are designed, between 5 and 50 parameters are typically extracted [14, 20, 21]. This large data set needs to be checked for technical errors and inconsistencies (Fig. 1). As a first quality control check, visualization of raw parameter values as pseudo-colored heat maps in multiwell plate layouts can reveal unintended batch, plate and plate position effects. We do this for one core parameter per fluorescent channel, e.g., neuron count per well, total dendrite length per neuron, presynaptic puncta density and postsynaptic puncta density. Plate position effects are easily observed as deviations from average in multiple adjacent wells in a particular area of the plate. Since randomized plate layouts are used in all our experiments (*see* Subheading 2.3), experimental manipulations usually do not result in strong plate position effects, and affected wells probably suffered from suboptimal culturing, staining or imaging conditions and should be removed from the analysis. Differences between plates or batches are acceptable to some extent and, unless extreme, are correctable via data normalization. In addition to visual inspection, the distribution of every parameter over all experimental and control wells of every plate separately is plotted. In a successful experiment, this should reveal little variation in negative control wells, while experimental wells will often show more variation due to experimental manipulation. In the case of neuronal network development, a time-dependent increase should be observed in dendrite length and synapse density from DIV7 to DIV21, while neuron numbers slightly decrease over time. Outliers, in particular in the negative control wells, should be manually inspected and removed if necessary.

3.2   Scaling and Normalization of Cellomics Data

All parameter values are log-transformed and normalized per plate by subtracting the average value of the negative controls. For this reason, it is strongly recommended to have at least 5 replicates of the negative controls on every plate, and to have removed all the negative control wells that showed outlier behavior (*see* Subheading 3.1). In RNAi screens, we prefer the use of scrambled shRNA samples for normalization, while using the untreated controls to subsequently determine the no-effect range of the shRNA treatment. Visual inspection of the normalized data is performed by rechecking parameter distributions as in Subheading 3.1.

3.3   Initial Exploration of Cellomics Data

Before any statistical analysis can be performed, parameters need to be checked for normality of distribution. Log-normalized data usually do not deviate much from a normal distribution, but in neuronal assays we often observe data distributions that are slightly skewed towards negative values because more treatments (RNAi or chemical) effect neuronal development negatively than positively. Parameters that show overall nonnormal distributions may need additional transformation steps before statistical analysis can be performed. Parameter correlation analysis provides information on parameter similarity and redundancy. Many HCS parameters are derived from higher order parameters or measure the same underlying biology in different ways. Parameters that have a correlation coefficient of >0.95 can be considered redundant. It is strongly recommended to remove redundant parameters or combine these (either weighed or un-weighed) into one single parameter in order to increase the power of subsequent statistical tests. Finally, we apply multilevel modeling to the data to estimate the contribution of various experimental and technical sources of variation to the overall variance of each parameter. Mixed-effect models are fitted for every parameter of interest using the R package lme4 [22] taking into consideration batch number, plate number, plate position (edge or non-edge), DIV and shRNA treatment as potential sources of variation. When the estimated contribution of technical sources of variation (e.g., plate, batch or plate position) exceed those of experimentally induced variation (e.g., shRNA treatment), data quality is probably poor due to technical or procedural errors. Results from this multilevel modeling step can be used to further guide outlier removal, data scaling and normalization procedures and parameter selection.

3.4 Statistical Analysis of Cellomics Data

Statistical analysis, for every parameter, is aimed to detect whether experimentally induced values differ significantly from the negative controls. A permutation analysis is performed per gene/parameter combination across DIV or for each DIV separately. Per gene/parameter combination, a nonparametric Mann-Whitney U test is performed under the null hypothesis that the distribution of negative controls and shRNA-treated wells are the same. For each gene/parameter pair a null distribution of 10,000 p-values is generated by randomly shuffling the negative control and shRNA labels within each plate and performing a Mann-Whitney U test on this permuted sample per gene. The observed p-value in the original sample is then compared with the distribution of null-p-values to obtain an empirical p-value, i.e., the percentage of null-p-values that is larger than the observed p-value. To account for multiple testing, a second permutation analysis is performed, where the minimum p-value for all gene tests per permuted sample is recorded instead. The observed null distribution is then compared with the distribution of minimum null-p-values. The resulting empirical p-value accounts for the number of tests as well as correlations between test statistics and can be interpreted as a corrected p-value. This unbiased way of statistical testing provides solid evidence for the most robust results from an RNAi screen. When significant hits are identified, it confirms the robustness of the screen and the validity of the findings, however, it obviously also results in many false negative findings. Therefore, we propose additional approaches to extract biologically meaningful data from RNAi screens.

3.5 Cluster Analysis of Cellomics Data and Hit Selection

In addition to statistical analysis, it is informative to cluster genes based on their biological phenotypes. As a first step we perform hierarchical clustering of all individual shRNAs on the basis of their selected parameter values. These combined parameter values determine the phenotypic change associated with any shRNA, and the more similar this phenotypic change is, the closer together shRNAs will cluster in a clustering dendrogram. This type of analysis can be used to (a) determine how many different phenotypes can be reliably detected based on the selected parameters, and (b) how many shRNAs targeting the same gene also produce the same phenotype. Based on these findings, genes can be selected that have an associated phenotype with a certain minimum number of shRNA, e.g., 3 out of 5, producing that phenotype. Statistical testing can now be performed for these genes and these shRNAs only, which will likely uncover many of the false negatives that were not identified by unbiased statistical testing in the previous step. Based on statistical testing and cluster analysis together, genes that produce cell biological effects of interest upon knockdown are selected for proteomics analysis.

3.6 Analysis of Proteomics Data

Proteomics data are acquired from replicate cultures for each shRNA separately and from scrambled and untreated controls. Data obtained by MS are analyzed with Spectronaut Pulsar software. For peptide and protein identification we use a spectral library built from untreated primary culture samples. All subsequent statistical analysis and data handling are done in R Studio. In brief, we set a QC value threshold of 10^{-2} (log10 scale) for peptide inclusion. For peptide identification, the maximum number of replicates in which a peptide is allowed to fail the quality threshold is set as the number of replicates minus one (within group fail), and the maximum number of sample groups in which a peptide is allowed to fail is set as the number of sample groups minus one (between group fail). Before further analysis, external protein contaminants (e.g., keratins) are excluded. Data are Loess normalized and then $\log_2$-transformed. For the selection of statistically significantly regulated proteins we use a false discovery rate (FDR) cutoff of 0.01. Samples with QC values higher than 10^{-2}, skewed data distribution or coefficients of variation >0.15 are excluded from further analysis. Primary neuron cultures usually result in >2000 identifiable and quantifiable proteins. A quick way to test whether shRNA treatment was effective is to cluster samples based on the abundance values of all identified proteins. Separate clustering of shRNA and control samples indicates the presence robust shRNA effects. However, relatively small shRNA effects may not be able to drive separate clustering of experimental and control groups. This type of clustering should therefore not be used as a pass or fail criterion in the proteomics data analysis workflow.

3.7 Proteomics Analysis: Comparing shRNA Level Effects

For all shRNAs that are included in the proteomics analysis the level of knockdown is determined with quantitative RT-PCR. Because the level of knockdown and the nature of the off-target effects may differ between shRNAs targeting the same gene, significantly regulated proteins are determined for each shRNA separately. The overlap between shRNAs is determined post hoc by comparing lists of significantly regulated proteins. We first set selection criteria for up- and downregulated proteins per shRNA based on log-fold change and p-value and then search for a minimal overlap between two or more shRNAs that produced the same phenotype. Optimal selection criteria differ per gene depending on differences in knockdown efficiency between shRNAs, the biological effect of the knockdown and the quality of the proteomics data. Sometimes it's helpful to apply more relaxed selection criteria per shRNA (i.e., lower log-fold change and/or higher p-value) while keeping a maximum overlap between shRNAs (i.e., 3 or more), whereas other times stringent selection criteria for fewer shRNAs may yield the highest overlap.

3.8 Proteomics Analysis: Comparing Gene Level Effects

The goal of comparing shRNA-level effects is to identify proteins whose levels are reliably and reproducibly changed due to decreased expression of a gene of interest. This reliable pool of regulated proteins is then tested for functional enrichment using bioinformatics tools (e.g., g-profiler, panther). Other resources may be consulted in particular instances. For instance, when the gene that was knocked down encodes a transcription factor or regulates transcription indirectly, it may be useful to search transcription factor binding site databases (e.g., TransFac, JASPAR) for the enrichment of binding motifs in the promoters of the genes corresponding to the set of regulated proteins. These types of analysis may identify biological processes or pathways that are associated with the knockdown of a particular disease risk gene. At this level of analysis, it is probably also interesting to compare different genes with respect to their dysregulated proteins and associated pathways. Protein lists can be merged in order to identify common dysregulated proteins that are shared between genes of interest as well as potential hub proteins that link overlapping protein sets and pathways.

3.9 Linking Proteomics Finding to Genetics and Disease

Genes in the proteomics analysis may share disease genetics, i.e., are potential risk genes for a certain disease, and may produce similar or overlapping neuronal morphological phenotypes upon knockdown. Networks of commonly dysregulated proteins between genes of interest may therefore hold valuable new information about the underlying biology of the disease, and it is credible that hubs within these shared protein networks may provide links to previously ignored risk genes or functional pathways.

Methods to uncover such links in a disease-relevant and statistically significant way are currently being developed. Lundby et al. [23] showed for instance that integration of a long QT syndrome (LQTS) protein–protein interaction network with LQTS GWAS data helps to filter weak GWAS signals and identify novel risk genes with nearby SNPs. Similar approaches have resulted in the identification of novel cancer risk genes [24, 25]. Improved genome-wide protein–protein networks have been generated with the specific purpose to provide better functional overlap with genetics data [26], and various tools have been developed to interrogate protein network and genetics data simultaneously, e.g., Gravity (gravity. pasteur.fr), GeNets and NetSig [27, 28]. Although primarily developed to work with protein–protein interaction networks, networks of dysregulated proteins upon knockdown of one or more specific genes, or predicted protein–protein interaction networks thereof, may be interrogated in the same way and provide valuable new information with respect to disease biology and genetics.

4 Conclusions

The study of disorders, such as SCZ, is challenging due to the many genes involved and the complexity of the underlying biology. We propose a combination of two powerful -omics approaches, cellomics, and proteomics, to investigate in medium-throughput and high-content mode the underlying biology of SCZ risk genes. We describe a data analysis pipeline for the selection of robust hits from RNAi screens based on in vitro high-content neuronal morphological data. Next we illustrate how protein expression changes associated with the knockdown of selected risk genes may reveal novel information about the biology of SCZ. Finally, we provide suggestions as to how proteomics data may be used to increase our understanding of the genetics of the disease and uncover previously unknown or ignored risk genes. The circular design of our approach (Fig. 1) provides translational value and clinical perspective to the results. Moreover, the approach can be easily adopted to the study of other disorders for which the essential biology can be modeled in vitro and robust cellular assays can be designed.

Acknowledgments

MR was supported by the European grant U-FP7 MC-ITN IN-SENS (#607616) and the Schizophrenia United Network (SUN) project. We gratefully acknowledge support from the Swedish Research Council (Vetenskapsrådet, award D0886501).

References

1. Sullivan PF, Daly MJ, O'Donovan M (2012) Genetic architectures of psychiatric disorders: the emerging picture and its implications. Nat Rev Genet 13(8):537–551

2. Gejman PV, Sanders AR, Duan J (2010) The role of genetics in the etiology of schizophrenia. Psychiatr Clin North Am 33(1):35–66

3. Schizophrenia Working Group of the Psychiatric Genomics Consortium (2014) Biological insights from 108 schizophrenia-associated genetic loci. Nature 511(7510): 421–427

4. Pardinas AF, Holmans P, Pocklington AJ, Escott-Price V, Ripke S, Carrera N, Legge SE, Bishop S, Cameron D, Hamshere ML et al (2018) Common schizophrenia alleles are enriched in mutation-intolerant genes and in regions under strong background selection. Nat Genet 50(3):381–389

5. Failla P, Romano C, Alberti A, Vasta A, Buono S, Castiglia L, Luciano D, Di Benedetto D, Fichera M, Galesi O (2007) Schizophrenia in a patient with subtelomeric duplication of chromosome 22q. Clin Genet 71(6):599–601

6. Fryland T, Christensen JH, Pallesen J, Mattheisen M, Palmfeldt J, Bak M, Grove J, Demontis D, Blechingberg J, Ooi HS et al (2016) Identification of the BRD1 interaction network and its impact on mental disorder risk. Genome Med 8(1):53

7. Huang CC, Cheng MC, Tsai HM, Lai CH, Chen CH (2014) Genetic analysis of GABRB3 at 15q12 as a candidate gene of schizophrenia. Psychiatr Genet 24(4):151–157

8. Stoll G, Pietilainen OPH, Linder B, Suvisaari J, Brosi C, Hennah W, Leppa V, Torniainen M, Ripatti S, Ala-Mello S et al (2013) Deletion of TOP3beta, a component of FMRP-containing mRNPs, contributes to neurodevelopmental disorders. Nat Neurosci 16(9):1228–1237

9. Vacic V, McCarthy S, Malhotra D, Murray F, Chou HH, Peoples A, Makarov V, Yoon S, Bhandari A, Corominas R et al (2011) Duplications of the neuropeptide receptor gene VIPR2 confer significant risk for schizophrenia. Nature 471(7339):499–503

10. Harrill JA, Robinette BL, Mundy WR (2011) Use of high content image analysis to detect chemical-induced changes in synaptogenesis in vitro. Toxicol In Vitro 25(1):368–387

11. Jain S, van Kesteren RE, Heutink P (2012) High content screening in neurodegenerative diseases. J Vis Exp (59):e3452

12. Kozak K (2009) Data mining techniques in high content screening: a survey. J Comput Sci Syst Biol 02(04)

13. Linhoff MW, Lauren J, Cassidy RM, Dobie FA, Takahashi H, Nygaard HB, Airaksinen MS, Strittmatter SM, Craig AM (2009) An unbiased expression screen for synaptogenic proteins identifies the LRRTM protein family as synaptic organizers. Neuron 61(5):734–749

14. Nieland TJ, Logan DJ, Saulnier J, Lam D, Johnson C, Root DE, Carpenter AE, Sabatini BL (2014) High content image analysis identifies novel regulators of synaptogenesis in a high-throughput RNAi screen of primary neurons. PLoS One 9(3):e91744

15. Sharma K, Choi SY, Zhang Y, Nieland TJ, Long S, Li M, Huganir RL (2013) High-throughput genetic screen for synaptogenic factors: identification of LRP6 as critical for excitatory synapse development. Cell Rep 5(5):1330–1341

16. Cohen E, Ivenshitz M, Amor-Baroukh V, Greenberger V, Segal M (2008) Determinants of spontaneous activity in networks of cultured hippocampus. Brain Res 1235:21–30

17. Chen N, Koopmans F, Gordon A, Paliukhovich I, Klaassen RV, van der Schors RC, Peles E, Verhage M, Smit AB, Li KW (2015) Interaction proteomics of canonical Caspr2 (CNTNAP2) reveals the presence of two Caspr2 isoforms with overlapping interactomes. Biochim Biophys Acta 1854(7):827–833

18. Pandya NJ, Klaassen RV, van der Schors RC, Slotman JA, Houtsmuller A, Smit AB, Li KW (2016) Group 1 metabotropic glutamate receptors 1 and 5 form a protein complex in mouse hippocampus and cortex. Proteomics 16(20):2698–2705

19. Pandya NJ, Koopmans F, Slotman JA, Paliukhovich I, Houtsmuller AB, Smit AB, Li KW (2017) Correlation profiling of brain subcellular proteomes reveals co-assembly of synaptic proteins and subcellular distribution. Sci Rep 7(1):12107

20. Daub A, Sharma P, Finkbeiner S (2009) High-content screening of primary neurons: ready for prime time. Curr Opin Neurobiol 19(5):537–543

21. Twarog NR, Low JA, Currier DG, Miller G, Chen T, Shelat AA (2016) Robust classification of small-molecule mechanism of action using a minimalist high-content microscopy screen and multidimensional phenotypic trajectory analysis. PLoS One 11(2):e0149439

22. Bates D, Mächler M, Bolker B, Walker S (2015) Fitting linear mixed-effects models using lme4. J Stat Softw 67(1)

23. Lundby A, Rossin EJ, Steffensen AB, Acha MR, Newton-Cheh C, Pfeufer A, Lynch SN,

Consortium QTIIG, Olesen SP, Brunak S et al (2014) Annotation of loci from genome-wide association studies using tissue-specific quantitative interaction proteomics. Nat Methods 11(8):868–874

24. Kamburov A, Lawrence MS, Polak P, Leshchiner I, Lage K, Golub TR, Lander ES, Getz G (2015) Comprehensive assessment of cancer missense mutation clustering in protein structures. Proc Natl Acad Sci U S A 112(40): E5486–E5495

25. Rosenbluh J, Mercer J, Shrestha Y, Oliver R, Tamayo P, Doench JG, Tirosh I, Piccioni F, Hartenian E, Horn H et al (2016) Genetic and proteomic interrogation of lower confidence candidate genes reveals signaling networks in beta-catenin-active cancers. Cell Syst 3(3):302–316.e4

26. Li T, Wernersson R, Hansen RB, Horn H, Mercer J, Slodkowicz G, Workman CT, Rigina O, Rapacki K, Staerfeldt HH et al (2017) A scored human protein-protein interaction network to catalyze genomic interpretation. Nat Methods 14(1):61–64

27. Horn H, Lawrence MS, Chouinard CR, Shrestha Y, Hu JX, Worstell E, Shea E, Ilic N, Kim E, Kamburov A et al (2018) NetSig: network-based discovery from cancer genomes. Nat Methods 15(1):61–66

28. Li T, Kim A, Rosenbluh J, Horn H, Greenfeld L, An D, Zimmer A, Liberzon A, Bistline J, Natoli T et al (2018) GeNets: a unified web platform for network-based genomic analyses. Nat Methods 15(7):543–546

INDEX

Printed by Printforce, the Netherlands